Pictorial Handbook of Medically Important Fungi and Aerobic Actinomycetes

Michael R. McGinnis

Associate Director, Clinical Microbiology
North Carolina Memorial Hospital
University of North Carolina at Chapel Hill
Chapel Hill, North Carolina

Richard F. D'Amato

Director, Division of Microbiology
The Catholic Medical Center of Brooklyn and Queens
Jamaica, New York

Geoffrey A. Land

Director, Department of Microbiology
Methodist Hospital of Dallas
Dallas, Texas

PRAEGER

PRAEGER SPECIAL STUDIES • PRAEGER SCIENTIFIC

Library of Congress Cataloging in Publication Data

McGinnis, Michael R.
  Pictorial handbook of medically important fungi and
aerobic actinomycetes.

  Bibliography: p.
  Includes index.
  1. Fungi, Pathogenic—Atlases. 2. Fungi, Pathogenic
—Identification. 3. Actinomycetaceae—Atlases.
4. Actinomycetaceae—Identification. I. D'Amato,
Richard F. II. Land, Geoffrey A. III. Title. [DNLM:
1. Actinomycetales—Laboratory manuals. 2. Fungi—
Laboratory manuals. QW 25 M145p]
QR245.M37   616'.015   81-5306
ISBN 0-03-058364-0 (pbk.)   AACR2

Published in 1982 by Praeger Publishers
CBS Educational and Professional Publishing
A Division of CBS, Inc.
521 Fifth Avenue, New York, New York 10175 U.S.A.

© 1982 by Praeger Publishers

All rights reserved

456789 052 9876543

Printed in the United States of America

Preface

During the past several years there have been many major advancements in the taxonomy of medically important aerobic actinomycetes and fungi, as well as in the techniques used for their identification. Although many scientific papers, reviews, monographs, and textbooks contain information on this subject, few have been adequately directed toward the clinical microbiologist, medical technologist, or medical technology student. Such works are often published in journals and books that are not commonly found in hospital laboratories, and they are frequently so expensive that many laboratories cannot afford to purchase them. Yet the staffs of clinical laboratories are the groups that need to increase their expertise and confidence in medical mycology.

The development of this pictorial handbook is the result of countless hours spent in preparing continuing education programs and workshops for laboratorians. We quickly recognized the frustration laboratorians experience in finding a single, simple, complete, and inexpensive reference work for the basic identification of the commonly encountered aerobic actinomycetes, moulds, and yeasts. This pictorial handbook was written to fill this void.

The organisms treated in this handbook were chosen on the basis of three criteria: their occurrence as common isolates from clinical specimens; their role as etiologic agents of infection in man, especially opportunistic infections; and their frequent occurrence as proficiency-testing unknowns. The handbook has been organized to permit laboratorians with a minimal mycological knowledge to identify most of the medically important aerobic actinomycetes and fungi. To assist laboratorians in achieving this goal, a brief introduction to some of the newer identification concepts and procedures and a glossary are included. The dichotomous word and pictorial keys have been designed to assist the laboratorian in achieving a rapid and accurate identification of an unknown organism. Because microbiologists and medical technologists often rely heavily upon photomicrographs for identifications, we have included photomicrographs that clearly and concisely illustrate the salient characteristics of each major organism. If an unknown organism is not compatible with the photomicrographs included in this handbook, then more extensive mycological monographs must be consulted.

The yeast identification section has been kept brief and basic. When dealing with the yeasts included in this work, only the assimilation reactions and morphologic characteristics included are necessary for their identification. Traditional fermentation reactions have been excluded because they require too much time before they can be read as negative. The substrates included in this handbook for the assimilation studies replace the need for fermentation reactions for only those yeasts in this handbook. Owing to the importance of *Cryptococcus neoformans,* several rapid methods for its tentative identification are outlined in the methods section.

We have prepared a simple but accurate identification scheme for the aerobic actinomycetes. A combination of morphologic and biochemical characteristics is necessary for the identification of the more common species of aerobic actinomycetes encountered in the clinical laboratory. Although thin-layer and gas-liquid chromatography appear to be the most definitive methods of identifying these bacteria, these techniques are not within the scope of most clinical laboratories. Therefore, we have elected not to use them. We use conventional physiological identification techniques for identifying members of this group of bacteria.

This handbook achieves two important goals in the diagnostic clinical laboratory. It provides a high quality, concise, and accurate pictorial reference for the identification of aerobic actinomycetes and fungi commonly encountered in the clinical laboratory. Second, it presents this information in a format that can be readily used by the clinical microbiologist, medical technologist, and medical technology student to gain the confidence necessary to implement a sound diagnostic mycology program.

Acknowledgments

We would like to thank Mrs. June Brown and Drs. Libero Ajello, Clete Kurtzman, and Arvind Padhye for their valuable comments and suggestions, which have greatly enhanced the quality of this book. We also wish to thank Ms. Christine Pinello, Drs. Libero Ajello, Arif El-Ani, and Arvind Padhye for permission to reproduce some of their excellent photomicrographs, and Mrs. Lynne Sigler for the loan of slide preparations maintained at the University of Alberta Mold Herbarium and Culture Collection.

Contents

Pictorial Handbook of Medically Important Fungi and Aerobic Actinomycetes

Introduction

Fungi are eukaryotic organisms with cells whose nuclei contain several chromosomes, a nucleolus, and a nuclear envelope that persists during nuclear division. The cell walls of fungi consist of repeating polysaccharide polymers of glucan, chitosans, mannan, and chitin. These components may be combined either with each other or with protein, lipids, or both. Bacteria such as the aerobic actinomycetes have cell walls containing peptidoglycans composed of polysaccharides other than those found in the fungi, lipid constituents, muramic acid, and diaminopimelic acid. Fungi lack chlorophyll, have mitochondria, an endoplasmic reticulum, 80S ribosomes, and are heterotrophic. They are usually distinguished from filamentous bacteria by being larger in size, resistant to antibacterial agents such as penicillin and streptomycin, and sensitive to antifungal agents such as amphotericin B and 5-fluorocytosine. In addition, fungi lack the ability to serve as hosts for bacteriophages.

Fungi grow vegetatively as yeasts, moulds, or a combination of both. Yeasts typically are recognized first by their moist, pasty colonies that consist of unicellular forms producing budded cells called blastoconidia (sing. blastoconidium). In contrast to the pasty colonies of yeasts, moulds form colonies that are usually velvety, woolly, or cottony in texture. This is because moulds form filaments called hyphae (sing. hypha) rather than unicellular budding cells. Some fungi are dimorphic; that is they can grow either as a yeast or as a mould, depending upon such factors as temperature and nutrition. A few fungi of medical interest are polymorphic; that is, they develop several different morphologic forms at one time.

Most yeasts produce blastoconidia. A blastoconidium is an asexual reproductive unit (propagule) that is formed by a blowing-out process with a subsequently appearing constriction where the conidium and the parent cell are attached to each other. Once a site on the cell wall of the parent cell has given rise to a blastoconidium, it apparently cannot give rise to additional conidia. Some yeasts form chains of blastoconidia that do not separate from each other. Such chains of blastoconidia are referred to as pseudohyphae (sing. pseudohypha), or pseudomycelium, depending upon the quantity being described. Since pseudohyphae consist of elongated blastoconidia, the cells are constricted at their point of attachment to each other.

When a yeast is able to utilize a compound for growth in the presence of oxygen, it is assimilating the compound. When a yeast utilizes a compound for growth in the absence of oxygen, the yeast is fermenting the substance, resulting in the production of carbon dioxide. In general, if a yeast can ferment a compound, it can assimilate it too. The converse is not necessarily true. A fermentation reaction must be read only for the production of gas, and not for a change in the pH of the growth medium. By using a battery of different substrates, a specific biogram for each species of yeast can be determined, which is then used in conjunction with the morphology of the yeast to distinguish it from similar yeasts.

Moulds are differentiated from each other almost solely upon the basis of their morphology. When a large number of hyphae are present, they are referred to as mycelium. Because mycelium may be singular or collective, the expression mycelia is not needed. As a hypha grows in length by linear elongation, septa (sing. septum) or crosswalls form in a centripetal manner from the hyphal cell walls. Thus, the walls of hyphae are parallel and are not constricted at the septa. Some of the zygomycetes form hyphae that contain only rare septa. These are usually described as sparsely septate, or incorrectly aseptate. Hyphae having septa are referred to as being septate. When spores and conidia germinate, they form germ tubes. A germ tube is the beginning of a new hypha; therefore, it is not constricted at its point of origin from the spore or conidium. This characteristic is used in yeast identification to distinguish germ tubes from elongated blastoconidia, since the latter appear to be constricted at their origin. Mycelium can be subdivided into three basic categories: vegetative mycelium, which penetrate the medium in order to absorb nutrients; aerial mycelium, which grow above the agar surface and contribute to the texture of the colony; and fertile mycelium, which give rise to reproductive propagules. These types of mycelium collectively form the colony.

Most fungi reproduce by sexual, asexual, or both means. A few fungi remain sterile and are classified in the order Mycelia Sterilia. Sexual reproduction involves the joining of haploid (1N) nuclei to form a diploid (2N) nucleus. The diploid nucleus undergoes meiosis, which typically gives rise to four haploid nuclei. When meiosis occurs in a saclike cell, and is followed by cleaving of the entire content of the saclike cell into haploid spores, the structure is called an ascus (pl. asci). The resulting spores are ascospores. Such development of sexual spores is characteristic of the Ascomycetes. In the Basidiomycetes, the process of meiosis occurs in a cell called a basidium (pl. basidia). The haploid basidiospores develop on the outside of the basidium. Some fungi, such as the Zygomycetes, produce large, thick-walled resting spores in which meiosis occurs upon maturation. These thick-walled spores are called zygospores.

Yeasts such as *Saccharomyces cerevisiae* produce asci that contain ascospores among their vegetative cells. These spores are easily recognized because they are acid-fast. Most filamentous Ascomycetes seen in the clinical laboratory develop their asci and ascospores within macroscopic fruiting bodies. The fruiting bodies of major medical interest are cleistothecia (sing. cleistothecium), perithecia (sing. perithecium), and ascostromata (sing. ascostroma). Cleistothecia are round, macroscopic structures consisting of loosely organized walls that enclose randomly dispersed asci. They typically lack a mouth, or opening called an ostiole. The ascospores escape through the loosely organized wall of the fruiting body, or are released when the cleistothecium ruptures. Perithecia tend to be flask shaped, have an ostiole, and contain asci in either a basal bush or in a lining within the fruiting body. The ascospores formed in a perithecium usually escape through the ostiole. In both the cleistothecium and perithecium, the fruiting body and asci develop concurrently. In the ascostroma, a cavity is formed in the fruiting body, and then the asci develop within it. These fruiting bodies often are black and carbonaceous.

Some fungi have specialized cells called conidiogenous cells, which give rise to conidia (sing. conidium). A conidium is an asexual reproductive propagule that is nonmotile, usually readily separates from its parent cell, and arises in any manner other than one involving a cleavage process. When a fungus produces conidia of two different sizes in the same manner, the larger is called a macroconidium, and the smaller is called a microconidium. In contrast to conidia, spores are reproductive propagules that arise following either mitosis or meiosis. When mitosis is involved, the propagule arises in a sporangium (pl. sporangia). A sporangium is a large saclike cell in which the entire content is cleaved into sporangiospores. A merosporangium is a special type of sporangium in which all of the sporangiospores are aligned in a single row. Some sporangia contain only a few sporangiospores; some others contain only one sporangiospore. Such sporangia are referred to as sporangiola (sing. sporangiolum). A sporangium is borne on a specialized hypha called a sporangiophore. The sporangium may surround a sterile, domelike, swollen area called a columella (pl. columellae), which is found at the apex of some sporangiophores. A number of Zygomycetes have rootlike structures called rhizoids that are associated with either the vegetative hyphae or with sporangiophores.

Conidia form by one of two principal methods: blastic or thallic. Blastic development means there is a noticeable enlargement of the young conidium before it is separated from its parent cell by a septum. In blastic development, only part of the parent cell becomes the conidium. Blastic conidium development may involve different cell-wall layers. Holoblastic development is characterized by the fact that all the cell-wall layers of the conidiogenous cell contribute to the formation of the conidium. For example, *Candida albicans* forms holoblastic blastoconidia. Enteroblastic conidium development involves only the inner cell-wall layer(s) of the conidiogenous cell wall; an example are the conidia formed by the phialides of *Phialophora verrucosa*.

Thallic development, which is the other principal method of conidium development, is characterized by the recognizable growth of the young conidium only after the conidium has been separated from its parent cell by a septum. In this case, the entire cell is converted into the conidium. This conversion often is followed by secondary growth and subsequent enlargement of the newly formed conidium. Thallic-arthric is a term that describes thallic conidium development in which the fertile hyphae fragment and disarticulate into conidia. Such conidia are called arthroconidia. The separation of the conidia occur at thickened septa called double septa. These are a special kind of septa that split through their centers to release the arthroconidia. The fragmentation of the hyphae of *Geotrichum candidum* into arthroconidia is an example of one type of thallic development.

The first conidium formed by a phialide is holoblastic, but all of the subsequent conidia are enteroblastic. In this type of conidiogenous cell, the conidia emerge from a single point at the apex of the phialide in a basipetal manner. Basipetal describes the condition of a chain of conidia in which the youngest conidium is at the base of the chain and the oldest is at the tip of the chain. In contrast, acropetal refers to chains having the youngest conidium at the tip and the oldest conidium at the base. Species of *Aspergillus* form their phialides on either a vesicle or a row of sterile cells called metulae (sing. metula). Metulae have been incorrectly referred to as sterigmata in the past. The term sterigmata should be restricted to the structures upon which some basidiospores develop. When the phialides arise directly from the vesicle, this is referred to as a uniseriate arrangement. When the phialides are on metulae, the arrangement is biseriate. Metulae are also produced by the majority of the *Penicillium* species. In the penicillia, the metulae give rise to the clusters of phialides. The metulae typically are formed at the apices of the branches of the conidiophore.

A conidiophore is a specialized hypha that bears either conidia or conidiogenous cells. It can be either determinate or proliferous in growth. Determinate refers to a conidiophore that stops growing prior to or at the time when a conidium forms at its apex. If new growth occurs during or after the period in which the terminal conidium develops, then the conidiophore is proliferous. Some fungi, such as *Sporothrix schenckii,* produce sympodial conidiophores. In this type of conidiophore development, there is a successive development of new subterminal areas of vegetative growth following the formation of each new terminal conidium. Once a terminal conidium forms, a new growing point develops below and to one side of it. The process is repeated, which results in a conidiophore that has the appearance of a series of bent knees (that is, geniculate). In some instances, the conidiophore will appear swollen rather than geniculate.

An annellide is another type of conidiogenous cell produced by some medically important fungi. The first conidium produced by an annellide is holoblastic in origin; each subsequent conidium is enteroblastic. Each conidium forms through the scar where the former conidium developed. Such development is percurrent. When the conidium separates from the annellide, a ring of cell wall material called an annellation remains at the apex of the annellide. Annellations are often difficult to see, and require an oil-immersion objective for observation. In contrast to phialides, annellides increase in length at their apices, become narrower in diameter at the apex, and have annellations.

During adverse conditions, fungi may form chlamydoconidia. Chlamydoconidia are thick-walled, thallic conidia that serve as a survival unit. They typically are released from their parent hypha by breakage, or lysis, of the hypha. The mature chlamydospores produced by *Candida albicans* on yeast morphology agar are not true chlamydoconidia; they are thick-walled vesicles. A vesicle is a swollen cell, not a reproductive propagule. Vesicles can occur either within the hyphae, or at the apices of some conidiophores or sporangiophores.

The medically important aerobic actinomycetes form branching filaments that are approximately 1 μm or less in diameter, some of which may be aerial. They are Gram-positive bacteria that are capable of growing at 25–37°C. Some microbiologists inappropriately refer to the filaments of the actinomycetes as hyphae. The term hyphae should be used to refer only to the filaments produced by fungi, and not to those produced by prokaryotic bacteria. The aerobic actinomycetes reproduce by developing spores, by the fragmentation of their filaments into coccoid or bacillary forms, or by a combination of both methods. When spores are present, they can be solitary, in clusters or chains, or formed in sporangia. The various genera are distinguished form each other primarily by their morphology, by the chemical composition of their cells and cell walls, and by specific biograms. The determination of cell-wall and whole-cell components such as diaminopimelic acid, arabinose, galactose, madurose, and mycolic acid has become increasingly important in identifying these organisms.

General References

Aerobic Actinomycetes

Goodfellow, M., G. H. Brownell, and J. A. Serrano. 1976. *Biology of the Nocardiae.* Academic Press, London. 517 pp.

Goodfellow, M. and D. E. Minnikin. 1977. Nocardioform bacteria. *Ann. Rev. Microbiol.* 31:159–80.

Kalakoutskii, L. V. and N. S. Agre. 1976. Comparative aspects of development and differentiation in actinomycetes. *Bacteriol. Rev.* 40:469–524.

Mishra, S. K., R. E. Gordon, and D. A. Barnett. 1980. Identification of nocardiae and streptomycetes of medical importance. *J. Clin. Microbiol.* 11:728–36.

Prauser, H. (ed.). 1970. *The Actinomycetales.* Gustav Fischer Verlag, Jena. 439 pp.

Fungi

Alexopoulos, C. J. and C. W. Mims. 1979. *Introductory Mycology.* 3rd. ed. John Wiley, New York. 632 pp.

Emmons, C. W., C. H. Binford, J. P. Utz, and K. J. Kwon-Chung. 1977. *Medical Mycology.* 3d. ed. Lea and Febiger, Philadelphia. 592 pp.

Kendrick, B. (ed.). 1979. *The Whole Fungus.* National Museum of Natural Sciences, Ottawa. Vols. 1 and 2. 793 pp.

Lodder, J. (ed.). 1970. *The Yeasts. A Taxonomic Study.* 2d. ed. North-Holland Publishing, Amsterdam. 1,385 pp.

McGinnis, M. R. 1980. *Laboratory Handbook of Medical Mycology.* Academic Press, New York. 661 pp.

Rippon, J. W. 1974. *Medical Mycology. The Pathogenic Fungi and the Pathogenic Actinomycetes.* W. B. Saunders, Philadelphia. 587 pp.

Identification

2

PART 1   Aerobic Actinomycetes

The genera of aerobic actinomycetes discussed in this handbook are those most frequently encountered in the clinical microbiology laboratory. Some authors have combined those species placed in the genera *Actinomadura, Nocardia,* and *Nocardiopsis* into an expanded concept for the genus *Nocardia.* We have elected to maintain *Actinomadura, Nocardia,* and *Nocardiopsis* as separate genera. Various identification schemes have been developed that utilize characteristics such as macroscopic and microscopic morphology, physiologic traits, and chromatographic analysis of whole-cell hydrolysates for sugars, lipids, and isomers of diaminopimelic acid. Although whole-cell analysis results in a more definitive identification, it requires time, personnel, and training that are beyond the scope of most diagnostic laboratories. The approach we have taken consists of using a limited number of biochemical tests that apply only to those isolates that are aerobic, Gram-positive, capable of growth at 25–35°C, and produce aerial and vegetative filaments consisting of branched elements approximately 1μm in diameter or less. Isolates that do not fit this description or cannot be satisfactorily identified by the methods in this handbook, should be sent to a reference laboratory.

Since the acid-fast property of species of *Nocardia* in culture is extremely variable, often depending on such factors as the growth medium and age of the culture, we have omitted the acid-fast stain from our identification scheme. This stain is still extremely valuable for the examination of clinical specimens when *Nocardia* is suspected. As a replacement for the acid-fast stain in the diagnostic laboratory, the lysozyme resistance test has been incorporated into our identification scheme. The most common species of *Nocardia* encountered in the laboratory are lysozyme resistant.

The macroscopic morphology of the aerobic actinomycetes is extremely variable; it is influenced by the chemical composition

TABLE 2.1 Physiologic Characteristics of Selected Aerobic Actinomycetes

Test	*Actinomadura madurae*	*Actinomadura pelletieri*	*Nocardia asteroides*	*Nocardia brasiliensis*	*Nocardia caviae*	*Nocardiopsis dassonvillei*	*Streptomyces griseus*	*Streptomyces somaliensis*	*Streptomyces species*[a]
Acid from cellobiose	+[b]	−	−	−	−	+	+	−	V
Hydrolysis of									
casein	+	+	−	+	−	+	+	+	+
hypoxanthine	+	V[d]	−	+	+	+	+	−	+
tyrosine	+	+	−	+	V	+	+	+	+
urea	−[c]	−	+	+	+	V	+	−	V
xanthine	−	−	−	−	+	+	+	−	V
Lysozyme resistance	−	−	+	+	+	−	−	−	V

[a] Includes *S. albus, S. lavendulae,* and *S. rimosus*
[b] ≥90% of strains positive
[c] ≥90% of strains negative
[d] Variable

Source: Modified from Mishra, Gordon, and Barnett. 1980. *J. Clin. Microbiol.* 11:728-36.

of the culture medium, as well as by temperature and duration of incubation. The macroscopic descriptions for all genera in this book are based on growth upon Sabouraud dextrose agar incubated at 30–35°C. These descriptions are only generalizations to aid in determining which genus may be present.

When an aerobic actinomycete is obtained in the laboratory, its purity must be ascertained. The colonies should be streaked for isolation on Columbia agar containing colistin and nalidixic acid. These antimicrobial agents will help eliminate bacterial contaminants. The battery of tests listed in Table 2.1 should then be set up according to instructions in Appendix A.

Actinomadura Lechevalier and Lechevalier, 1970.

Description: Colonies are slow growing, aerobic, Gram-positive, nonacid-fast, glabrous, heaped and folded, and red, pink, yellow, orange, white, or tan in color. Filaments are branched with sparse to abundant aerial filaments that form chains of arthrospores. See Table 2.1 for physiologic characteristics.

Salient characteristics: Isolates of *Actinomadura* are susceptible to lysozyme and do not produce urease. *Actinomadura madurae* is distinguished from *A. pelletieri* by the fact that it produces acid from cellobiose.

Laboratory precautions: Handle with care, but special precautions are not necessary.

Key reference:
Sykes and Skinner, 1973

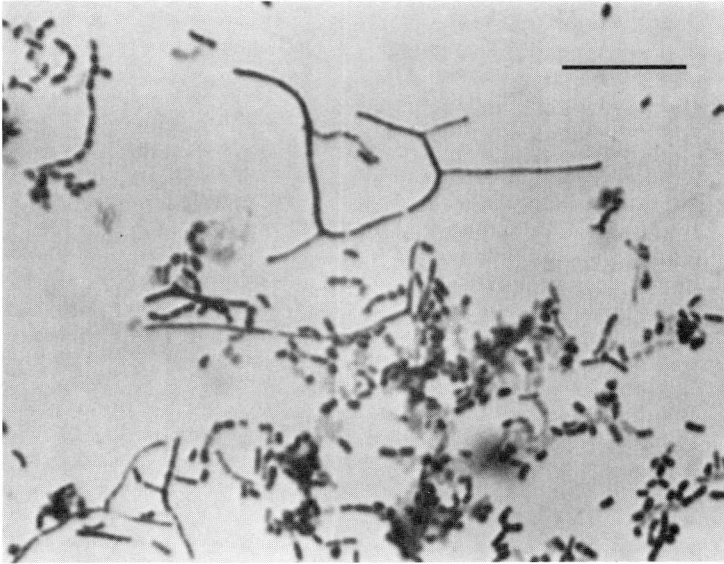

Figure 1. *Nocardia asteroides.* The branching filaments are Gram-positive. Bar is 10 μm.

Nocardia Trevisan, 1889.

Description: Colonies are slow growing, aerobic, Gram-positive, acid-fast to partially acid-fast, glabrous, heaped and folded, and white, pink, red, orange, or tan in color. Filaments are branched, fragmenting into rod and coccoid forms. Aerial filaments usually present. See Table 2.1 for physiologic characteristics.

Salient characteristics: Medically important species of *Nocardia* are lysozyme resistant and do not produce acid from cellobiose.

Laboratory precautions: Handle with care, but special precautions are not necessary.

Key references:
Goodfellow and Minnikin, 1977
Mishra, Gordon, and Barnett, 1980

Nocardiopsis Meyer, 1976

Description: Colonies are slow growing, aerobic, Gram-positive, nonacid-fast, coarsely wrinkled to folded, and white or pink to red in color. Filaments are well developed, branched, long, and fragmenting into spores. Aerial filaments are well developed and abundant. See Table 2.1 for physiological characteristics.

Salient characteristics: *Nocardiopsis dassonvillei* produces acid from cellobiose, hydrolyzes xanthine, and is sensitive to lysozyme.

Laboratory precautions: Handle with care, but special precautions are not necessary.

Key reference:
Meyer, 1976

Streptomyces Waksman and Henrici, 1943

Description: Colonies are slow growing, aerobic, Gram-positive, nonacid-fast, glabrous or chalky, heaped and folded, and white, tan, gray, brown, or black in color. Colonies often have an earthy odor. Filaments are extensively branched; aerial filaments are abundant and usually produce long chains of spores formed by fragmentation of the filaments. See Table 2.1 for physiologic characteristics.

Salient characteristics: Isolates of *Streptomyces* are frequently recognized by their chalky appearance and earthy odor.

Laboratory precautions: Handle with care, but special precautions are not necessary

Key references:
Kurylowicz *et al.,* 1975
Shirling and Gottlieb, 1972

PART 2   Fungi

It should be assumed that every culture is contaminated until proven differently. Once the unknown fungus has been streaked for colony isolation on media such as brain-heart infusion agar (yeasts) or potato dextrose agar (moulds) with or without antibacterial agents, it can be identified. Moulds should be studied in slide culture unless a teased mount reveals the diagnostic characteristics. *It is extremely important that all unidentified moulds be worked with in a biological safety cabinet.*

As an aid in identiyfing fungi, two keys are included in this portion of the handbook. The first is a traditional dichotomous key that includes yeasts and moulds. In using this key, one should read all sets of statements, starting with number 1. Select the description that best describes what is observed about the unknown fungus and then follow the leader dots to the right margin. Either a second number or a possible name for the fungus will be found. If a number is at the right margin, go to that number in this key and read the new set of descriptions. The reader is led through a series of elimination steps to a potential final identification. If the unknown fungus does not completely fit one of the choices in the pair of descriptions, the fungus must be keyed out as if both choices were correct. The user of a dichotomous key must always remember that the unknown fungus may not be included in the key. Once a potential identification is made, the unknown fungus should be compared and contrasted to the descriptions and photomicrographs in this handbook.

The second key included in this portion of the handbook is a pictorial key for identifying some of the medically important moulds. It is located in Section A.

Dichotomous Key to Some of the Genera of Medically Important Fungi[a]

1. Filaments 0.5–1.0 μm in diameter present; yeast cells and mycelium absent; sensitive to antibacterial agents; colonies waxy to chalky, often brightly colored, flat or raised, smooth or irregular in texture .... Actinomycetes (Part 1)
1. Filaments 0.5–0.1 μm in diameter absent; yeast cells, mycelium, or both present; sensitive to antifungal agents; colonies pasty to cottony in texture 2
    2. Budding yeast cells predominant; colonies white to black, moist, usually pasty in texture 5
    2.´ Mycelium predominant, yeast cells typically absent; colonies white to black, velvety to cottony in texture 3
3. Zygospores occasionally present; merosporangia, sporangia, or sporangiola typically present; mycelium usually sparsely septate and broad in diameter............. 16
3.´ Zygospores, merosporangia, sporangia, or sporangiola absent; mycelium septate and narrow in diameter 4
    4. Asci and ascospores present 23
    4.´ Asci and ascospores absent........................ 27
5. Colonies brown to black 6
5.´ Colonies white to brightly colored 9
    6. Colony color due to thick-walled, darkly pigmented arthroconidia .... see *Aureobasidium*
    6.´ Colony color due to dematiaceous budding yeast cells 7
7. Yeast cells 2-celled, elliptic, tapering toward one end .... see *Exophiala*
7.´ Yeast cells 1-celled, globose to elongate 8
    8. Yeast cells globose to ellipsoid, growing on the agar surface; hyphae absent; pseudohyphae, chains of globose cells, or both may be present .... *Phaeococcomyces* (also see *Exophiala* and *Wangiella*)
    8´ Yeast cells globose to elongate, commonly growing within the medium rather than on its surface; hyphae and pseudohyphae often present, hyphae may have clamp connections; teliospores and oval-to-crescent-shaped basidiospores may be present......*Ustilago*
9. Conidia forcibly discharged; colonies usually orange .... *Sporobolomyces*
9. Conidia not forcibly discharged.......................... 10
    10. Hyphae and pseudohyphae absent, if pseudohyphae present, then rudimentary.................... 11
    10.´ Hyphae, pseudohyphae, or both present............. 15

---

[a]Colony characteristics and microscopic descriptions for the moulds are based upon isolates growing on potato dextrose agar or cornmeal agar incubated for two weeks at 25°C in the dark. Characteristics for the yeasts are based upon physiologic tests and their morphology after four to five days at 25°C on Dalmau plates (cornmeal agar).

11.   Asci and ascospores present . . . . *Saccharomyces*
11.′ Asci and ascospores absent 12
       12.   Reproduction by unipolar budding; cells bottle-
             shaped . . . . *Malassezia*
       12′   Reproduction other than unipolar budding 13
13.   Urease formed 14
13.′ Urease not formed . . . . *Torulopsis* (also see *Saccharomyces*)
       14.   Inositol assimilated . . . . *Cryptococcus*
       14.′ Inositol not assimilated . . . . *Rhodotorula*
15.   Arthroconidia present . . . . *Trichosporon*
15.′ Arthroconidia absent . . . . *Candida*
       16.   Spores forcibly discharged 17
       16′   Spores not forcibly discharged 18
17.   Zygospores present, with beaklike structures from the
      suspensor cells . . . . *Basidiobolus*
17.′ Zygospores may be absent, if they are present, then
      without beaklike structures from the suspensor cells . . . . *Conidiobolus*
       18.   Merosporangia present . . . . *Syncephalastrum*
       18.′ Merosporangia absent 19
19.   Spores formed around vesicle at the apex of the
      sporangiophore . . . . *Cunninghamella*
19.′ Sporangia having many spores 20
       20.   Rhizoids formed opposite sporangiophores;
             sporangiophores unbranched and often in groups . . . . . *Rhizopus*
       20.′ Rhizoids, if present, not opposite sporangiophores 21
21.   Swelling in the sporangiophore where it merges with the
      columella; sporangium pear-shaped; septum typically in
      sporangiophore just below columella; sporangiophore
      branched . . . . *Absidia*
21.′ Swelling in sporangiophore where it merges with col-
      umella absent; sporangium typically round 22
       22.   Rudimentary rhizoids present; thermotolerant . . . . *Rhizomucor*
       22.′ Rhizoids absent; not thermotolerant . . . . *Mucor*
23.   Cleistothecia present (frequently within the agar),
      globose, without an ostiole, light brown to black; asci
      globose to subglobose with 8 ascospores; ascospores oval
      to ellipsoid, pale yellow-brown to copper, smooth,
      1-celled . . . . *Pseudallescheria*
23.′ Cleistothecia absent 24
       24.   Perithecia present, round to flask shaped, with os-
             tiole, dark brown to black, with long hairlike,
             brown to black appendages; asci clavate to cylin-
             drical with 8 ascospores; ascospores lemon shaped,
             usually olive brown, 1-celled . . . . *Chaetomium*
       24.   Perithecia absent · 25
25.   Ascospores 1-celled, hyaline to dark, fusoid, usually
      curved, with a tail; ascostromata nearly round to ir-
      regular; asci ellipsoid with 8 ascospores . . . . *Piedraia*
25.′ Ascospores 2- or more celled 26
       26.   Ascospores 4- to 9-celled, hyaline to dark, fusoid,
             curved, with a constriction at each septum;

ascostromata globose to nearly globose; asci clavate to cylindrical with 8 ascospores.................*Leptosphaeria*

  26.´ Ascospores 2-celled, brown to black, ellipsoidal or variable, central septum sharply constricted; ascostromata round to ellipsoid, black; asci globose to clavate with 8 ascospores.............................*Neotestudina*

27. Pycnidia absent.. 29

27.´ Pycnidia present 28

  28. Setae absent; pycnidia round to lens shaped, black, with ostiole; conidia globose to cylindrical, 1-celled, hyaline, usually with 2 oil droplets .... *Phoma*

  28.´ Setae present; pycnidia round to lens shaped, black, with ostiole; conidia globose to cylindrical, 1-celled, hyaline ·*Pyrenochaeta*

29. Synnemata present, dark, erect, single or in clusters; conidia subglobose to ovoid, 1-celled, hyaline, forming a slimy ball at the apex of the synnema; conidiogenous cells are annellides · *Graphium*

29.´ Synnemata absent · 30

  30. Colonies sterile · 31

  30.´ Colonies producing conidia · · · · · · · · · · · · · · · · · · 32

31. Colonies black, originating from cases of black grain mycetoma.. *Madurella*

31.´ Colonies white to pale colored; dimorphic; yeast form reproduces by multiple budding............................*Paracoccidioides*

  32. Colonies, hyphae, or conidia brown to black 33

  32.´ Colonies, hyphae, and conidia other than brown to black 51

33. Conidia more than 1-celled 34

33.´ Conidia 1-celled 41

  34. Conidia having vertical and transverse septa present 35

  34.´ Conidia having both vertical and transverse septa absent.. 38

35. Conidia in simple or branched chains; conidia typically rounded at their base, tapering toward their apices, forming a beak...................................... *Alternaria*

35.´ Conidia not occurring in chains...........................36

  36. Conidia developing on sporodochia; conidia rough with a warty crust, subglobose; colonies yellow to orange.................................... *Epiccocum*

  36.´ Sporodochia absent 37

37. Conidiophores septate, brown, sympodial, geniculate; conidia without central constriction ....*Ulocladium*

37.´ Conidiophores septate with swollen terminal portion, percurrent; conidia typically with a central constriction ..... *Stemphylium*

  38. Blastoconidia 1- to several-celled with dark hila, smooth or echinulate, occurring in fragile branched chains; conidiophores erect, pigmented .... *Cladosporium*

  38.´ Conidia not occurring in chains 39

39. Conidiophores septate, brown, sympodial, geniculate 40

39.´ Conidiophores septate, brown, with parallel walls; conidia multicelled, obclavate .... *Helminthosporium*

    40. Conidia curved with central cell typically larger than others, pale brown with end cells usually paler in color than central cells .... *Curvularia*

    40.´ Conidia cylindrical, darkly pigmented .... *Drechslera*

41. Conidia in chains 42

41.´ Conidia solitary or in balls 43

    42. Conidia having dark hila, smooth or echinulate; conidia occurring in fragile branched chains .... *Cladosporium*

    42.´ Conidia lacking hila, smooth; conidia occurring in series that resemble branched chains .... see *Fonsecaea*

43. Conidia not in balls 44

43.´ Conidia in balls 48

    44. Conidiophores hyaline, long, sympodial, often with a swollen apex; conidia frequently of 2 types: conidia of the first type are hyaline and arise from denticles along hyphae or conidiophores, second type of conidia dark colored, thick walled, arising along hyphae .... *Sporothrix*

    44.´ Characters not as above 45

45. Conidia solitary, black, smooth, globose to ovoid, horizontally flattened; conidiophores hyaline, swollen in the middle, tapering toward the point of attachment of the conidium .... *Nigrospora*

45.´ Conidia not horizontally flattened, not jet black; conidiophores unswollen, not tapering toward their apices 46

    46. Conidia pale brown to dark brown, smooth, globose to elongate, solitary, developing on denticles or in clusters from annellides .... *Scedosporium*

    46.´ Conidia hyaline to pale brown, along hyphae or as complex heads 47

47. Conidia hyaline, developing along hyphae that may be hyaline or pale brown; conidia commonly producing secondary 1-celled blastoconidia; 1- to 2-celled darkly pigmented arthroconidia commonly present .... *Aureobasidium*

47.´ Conidia pale brown, developing at apices of distinct sympodial conidiophores; primary conidia functioning as sympodial conidiogenous cells producing 1-celled secondary conidia; the process is repeated, resulting in a complex head consisting of series of conidia and conidiogenous cells; *Cladosporium*-like, *Phialophora*-like, or *Rhinocladiella*-like forms often present .... *Fonsecaea*

    48. Clusters of hyaline conidia along hyphae; hyphae hyaline to pale brown; conidia commonly producing secondary 1-celled blastoconidia; 1- to 2-celled darkly pigmented arthroconidia commonly present .... *Aureobasidium*

    48.´ Conidiogenous cells cylindrical to flask shaped 49

49. Conidiogenous cells annellides, pale brown, usually on distinct conidiophores, but may be integrated within the hyphae; yeast cells often present .... *Exophiala*

49.´ Conidiogenous cells phialides 50

63.′   Penicillus bearing a single large ball of conidia is absent.....64

    64.   Phialides occurring in whorls at the apices of ver-
            ticillately branched conidiophores .... *Verticillium*

    64.′   Phialides ovoid to flask shaped, swollen at the cen-
            tral portion, tapering toward the apex, solitary or
            in clusters; conidia 1-celled and hyaline to green;
            conidiophores at wide angles to hyphae; colonies
            rapid growing, at first white, later developing green
            tufts... *Trichoderma*

65.   Multicelled conidia present................................. 66

65.′   Multicelled conidia absent 68

    66.   Macroconidia typically spindle shaped, echinulate,
            with thickened cell walls; microconidia often pres-
            ent, 1-celled, smooth, clavate .... *Microsporum*

    66.′   Macroconidia cylindrical, smooth; microconidia
            may be present or absent 67

67.   Conidia 3- to 5-celled, club shaped, smooth, solitary or
    in clusters, thick walled; microconidia absent .... *Epidermophyton*

67.′   Conidia smooth, cylindrical to clavate, thin walled;
    microconidia typically present............................... *Trichophyton*

    68.   Conidia 1-celled, thick walled, golden brown, trun-
            cate with an annular frill, solitary, forming on a
            broad denticle; clamp connections typically present... *Sporotrichum*

    68.′   Conidia not thick walled and golden brown; clamp
            connections absent 69

69.   Conidiophores sympodial 70

69.′   Conidiophores, when present, not sympodial 71

    70.   Conidiophores flask shaped, tapering to form a
            zigzag-appearing rachis, hyaline, occurring in dense
            clusters along the hyphae; conidia 1-celled, hyaline,
            globose to ovoid, occurring along the rachis .... *Beauveria*

    70.′   Conidiophores hyaline, often with swollen apices;
            conidia frequently of 2 types: conidia of the first
            type hyaline, arising from denticles along the
            hyphae or conidiophores; conidia of the second
            type dark colored, thick walled, arising along the
            hyphae only .... *Sporothrix*

71.   Conidia large, thick walled 72

71.′   Conidia small to large, thin walled 73

    72.   Conidiophores hyaline, short or long; conidia ter-
            minal, solitary or in clusters, 1-celled, globose to
            ovoid, hyaline to amber, smooth or rough;
            microconidia absent .... *Sepedonium*

    72.′   Conidiophores hyaline, short, arising at 90°-angle
            to hyphae; conidia terminal, solitary, 8–14 μm,
            1-celled, globose, hyaline to amber, smooth to
            tuberculate, often with fingerlike projections;
            microconidia are hyaline, 2–4 μm, smooth or
            echinulate; dimorphic............................... *Histoplasma*

73.   Conidia 1-celled, hyaline, smooth to tuberculate, formed
    along the hyphae, on pedicels or on long branches; ran-

domly dispersed arthroconidia often abundant; solitary conidia and arthroconidia broader than hyphae..............*Chrysosporium* (also see *Paracocci-dioides*)

73.′   Conidia along hyphae or on short conidiophores; arthro-conidia absent...74

      74.   Conidia 1-celled, hyaline, solitary, smooth, globose to subglobose, forming on conidiophores that arise at a 90°-angle to the hyphae; dimorphic ....*Blastomyces*

      74.′   Conidia 1- to several-celled, globose- to- clavate-to-balloonlike, along the hyphae or in clusters; macroconidia may be present..........................*Trichophyton*

SECTION A Identification of Moulds

For those microbiologists who do not wish to use a dichotomous key, we have included a pictorial key. If an illustration in the key and the unknown fungus appear to be the same, the fungus should be compared and contrasted to the descriptions and photomicrographs included in this handbook. If the unknown fungus does not match one of the illustrations in a reasonable manner, the fungus is probably not included in this handbook.

The descriptions contain information regarding both the macroscopic and microscopic characteristics that are distinctive for each fungus. In those cases in which two or more fungi could be readily confused, the salient characteristics used to distinguish them have been included. Due to the pictorial nature of this handbook, the descriptions and key references are brief. The majority of the descriptions are based upon two-week-old cultures that were grown on potato dextrose agar at 25°C in the dark. The colonial descriptions for the species included in the genera *Epidermophyton, Microsporum,* and *Trichophyton* are based upon isolates on Sabouraud dextrose agar with 4% glucose; descriptions for *Aspergillus* species are based upon isolates on Czapek-Dox solution agar.

Bars for line drawings are 10 μm.

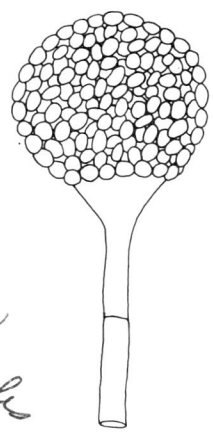
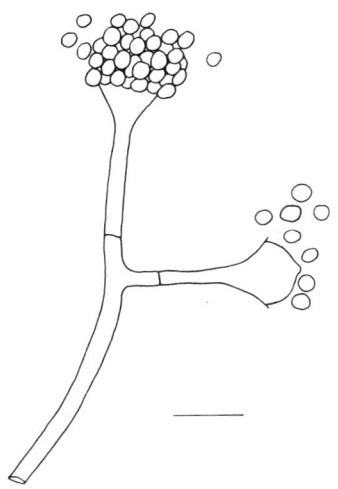

Absidia sp.

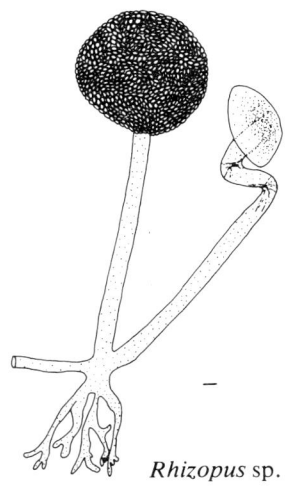

Rhizopus sp.

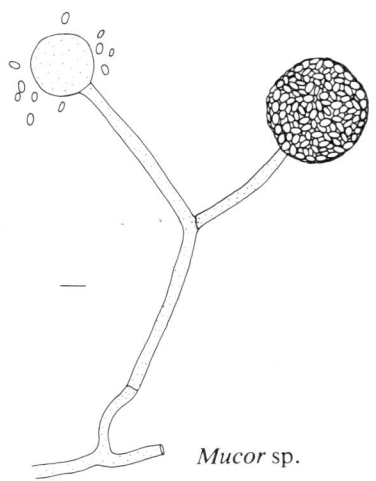

Mucor sp.

Rhizomucor pusillus

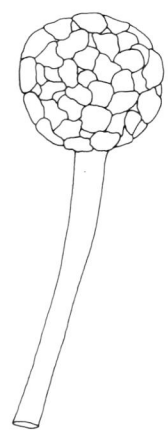
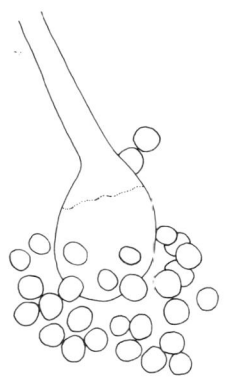

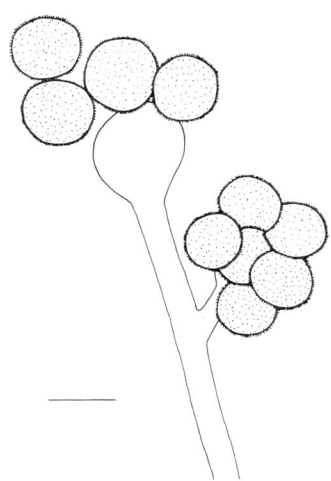

Cunninghamella bertholletiae

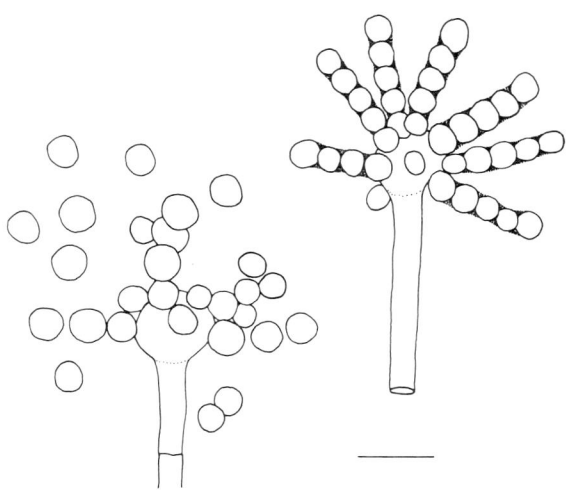

Syncephalastrum racemosum

21

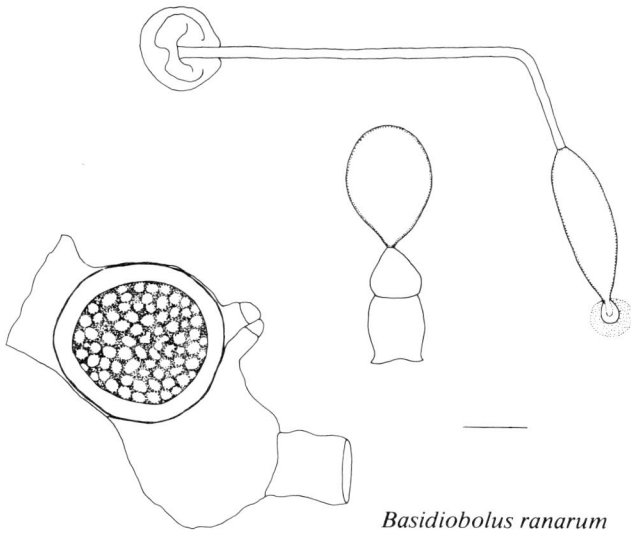

Basidiobolus ranarum

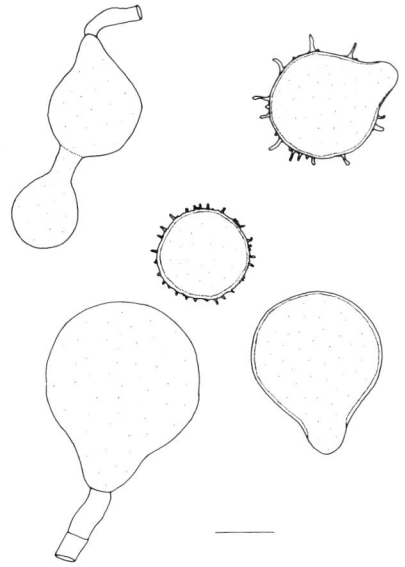

Conidiobolus coronatus

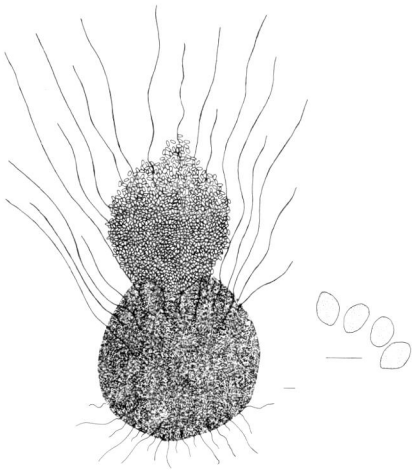

Chaetomium sp.

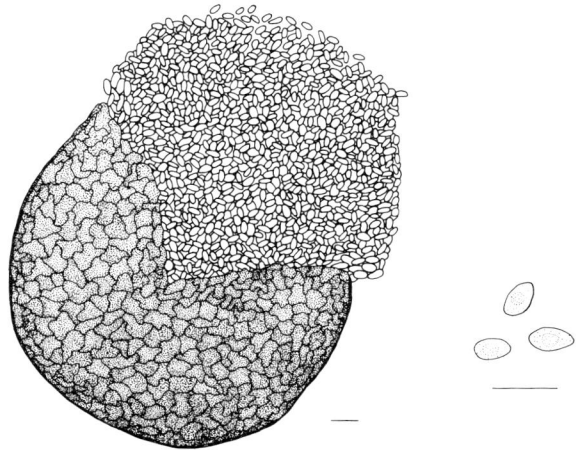

Pseudallescheria boydii

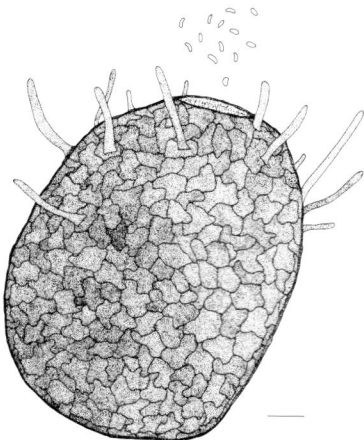

Pyrenochaeta romeroi

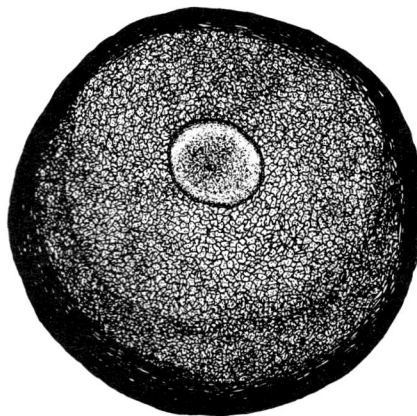

Phoma sp.

22

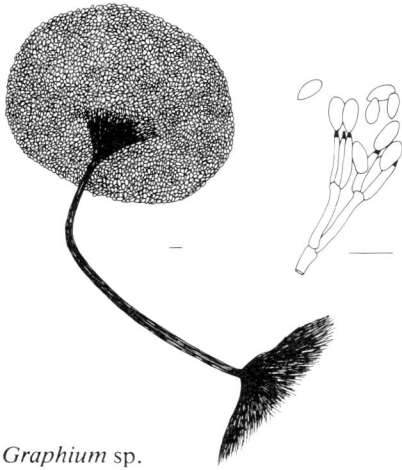

Graphium sp.

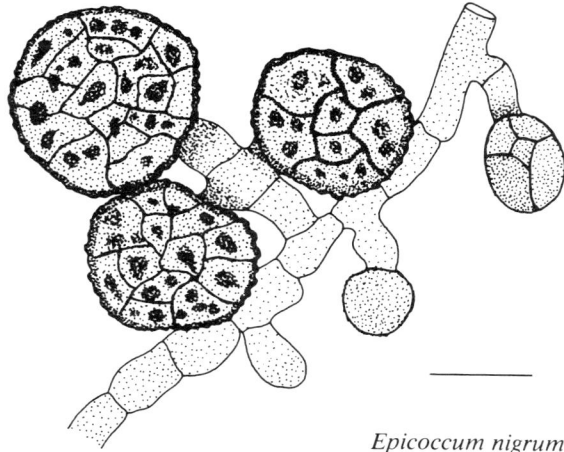

Epicoccum nigrum

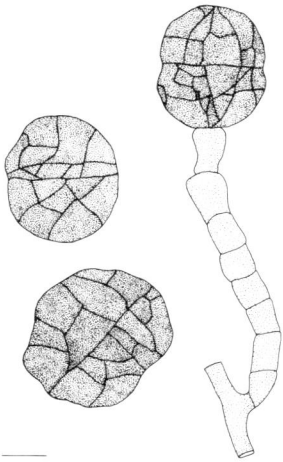

Stemphylium sarcinaeforme

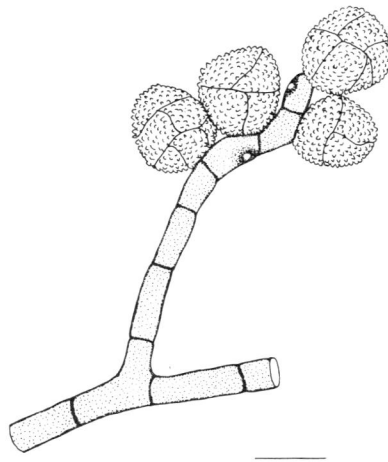

Ulocladium atrum

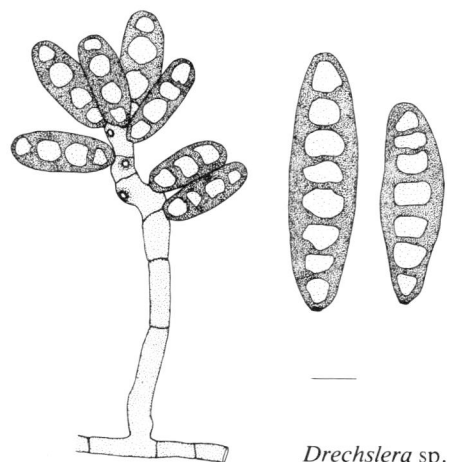

Drechslera sp.

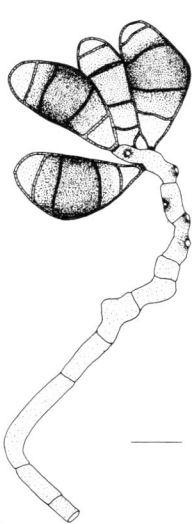

Curvularia sp.

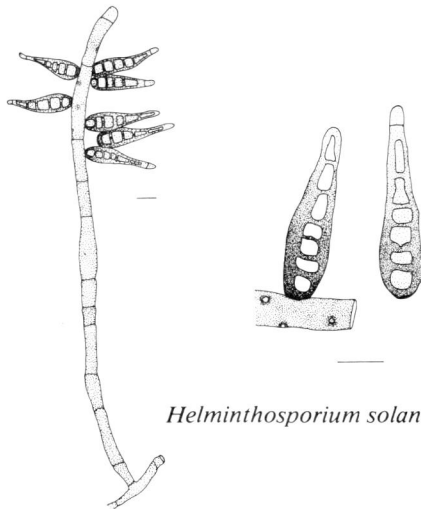

Helminthosporium solani

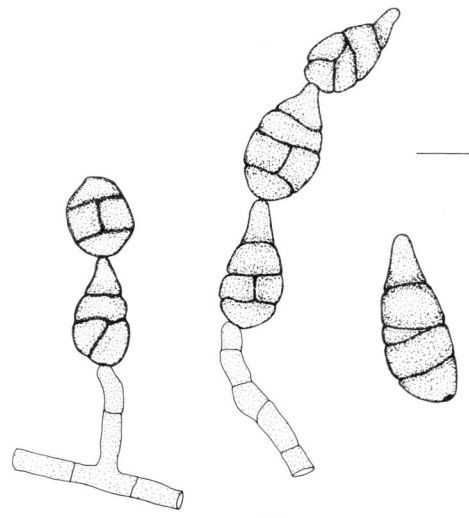

Alternaria sp.

23

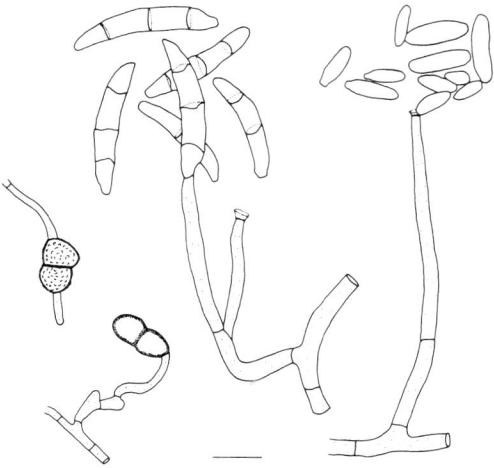

Fusarium solani

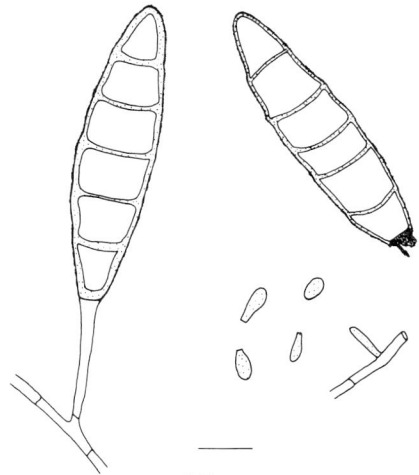

Microsporum gypseum

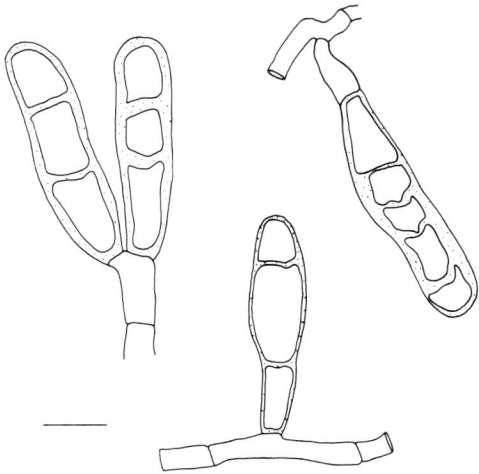

Epidermophyton floccosum

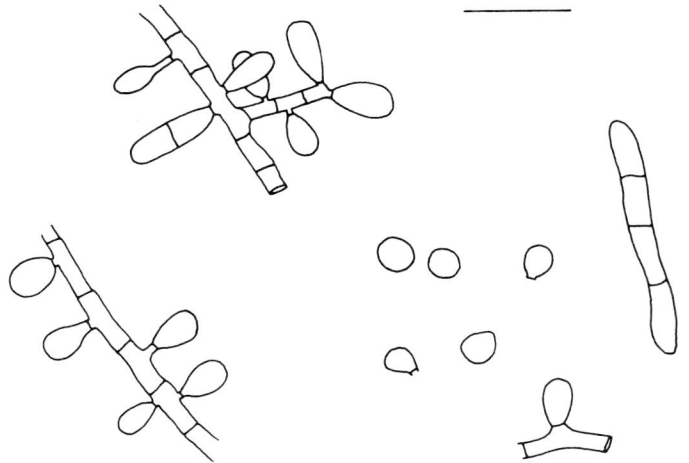

Trichophyton tonsurans

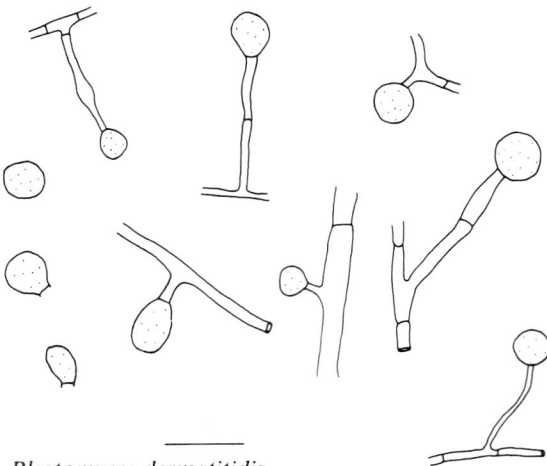

Blastomyces dermatitidis

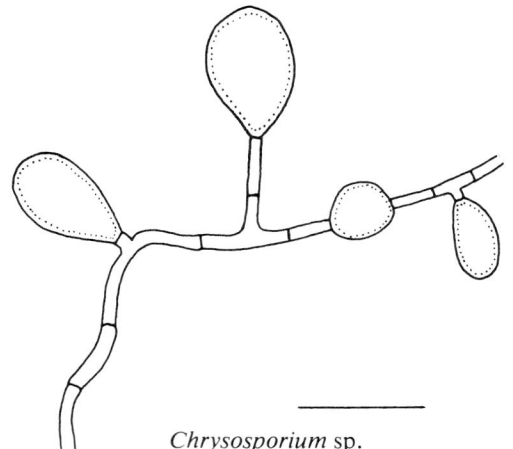

Chrysosporium sp.

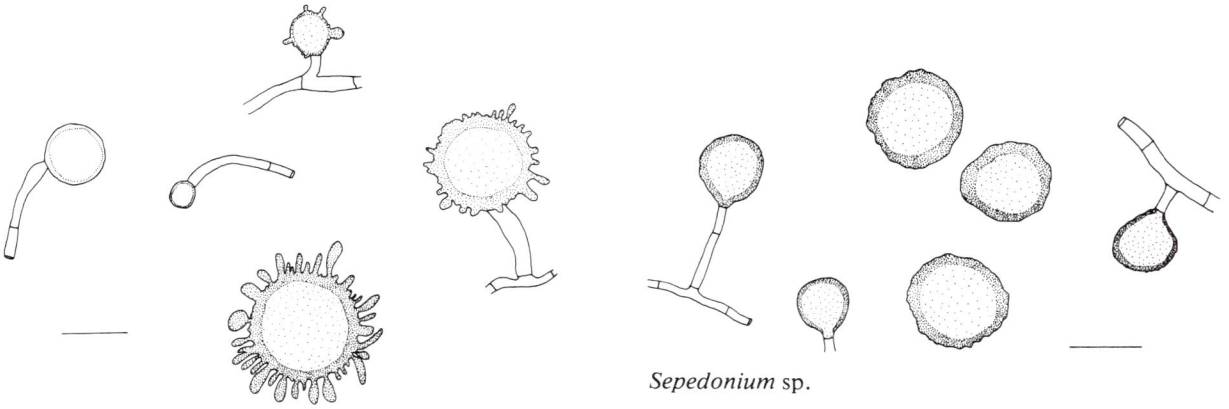

Histoplasma capsulatum

Sepedonium sp.

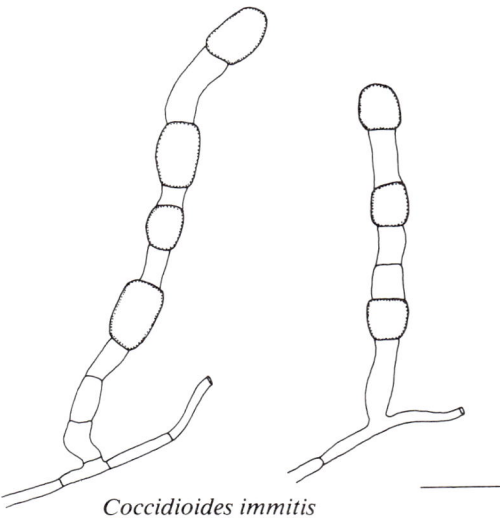

Coccidioides immitis

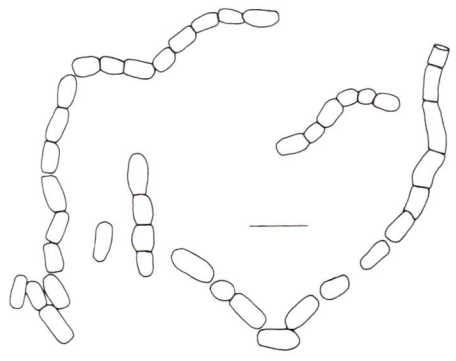

Geotrichum candidum

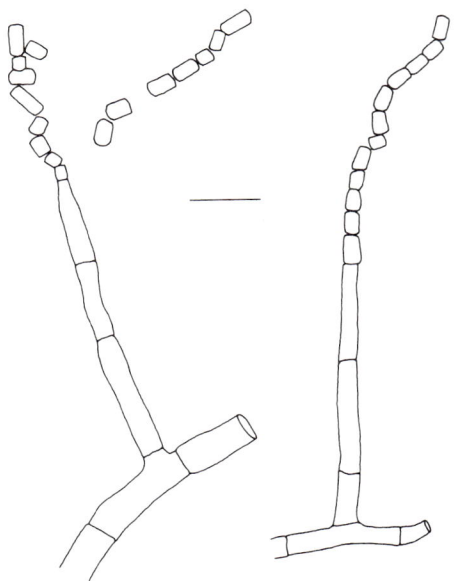

Arthrographis cuboidea

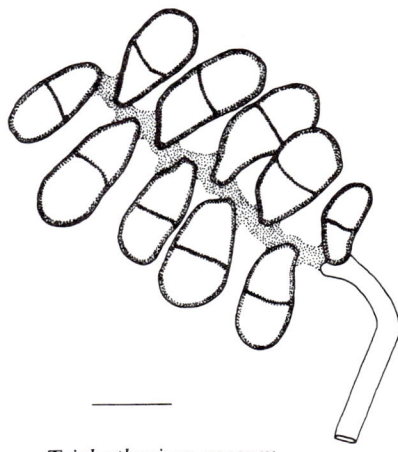

Trichothecium roseum

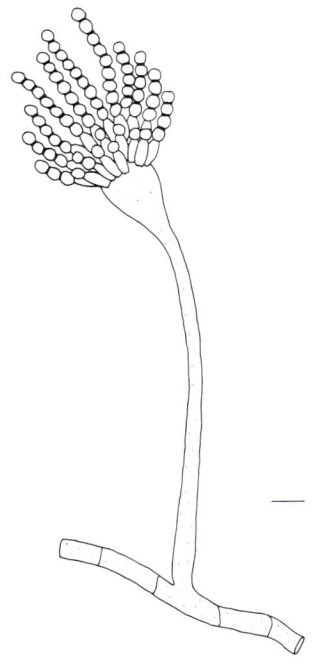

Aspergillus fumigatus

25

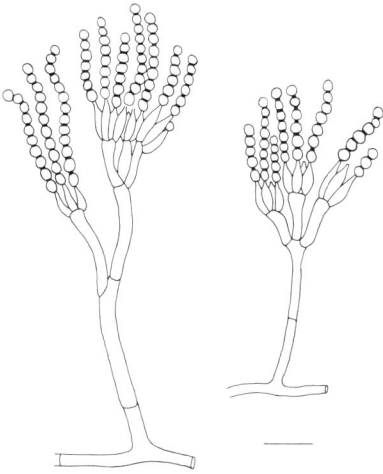

Penicillium sp.

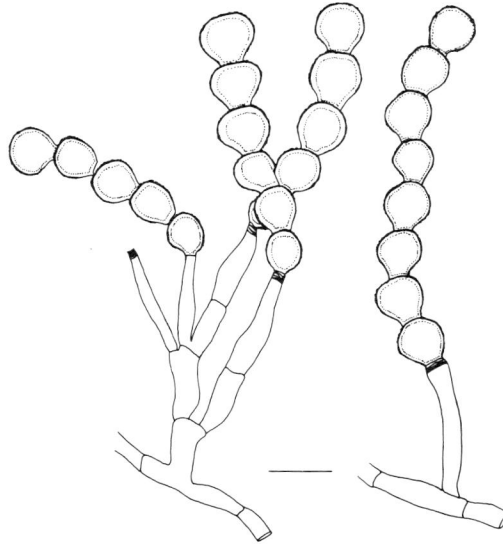

Scopulariopsis brevicaulis

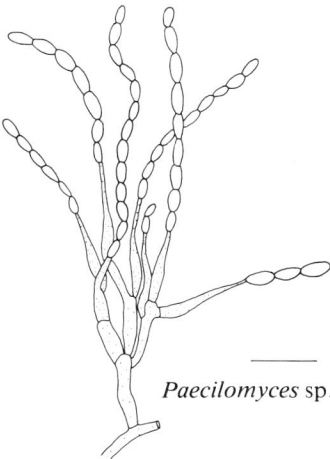

Paecilomyces sp.

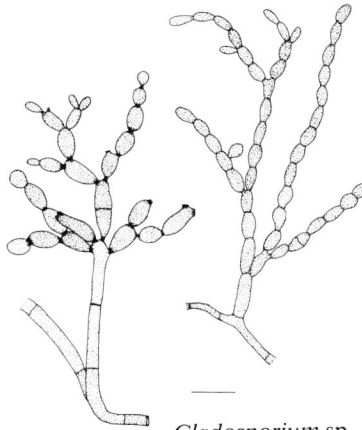

Cladosporium sp.

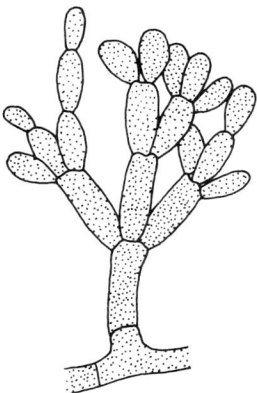

Fonsecaea pedrosoi

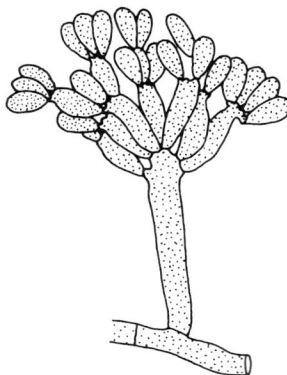

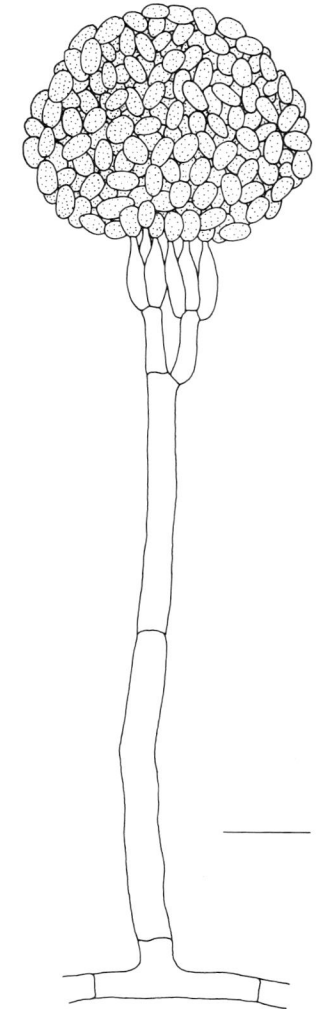

Gliocladium sp.

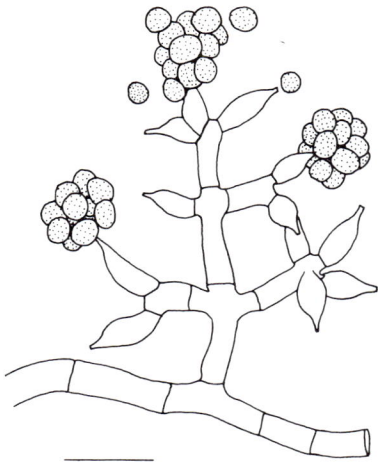

Trichoderma sp.

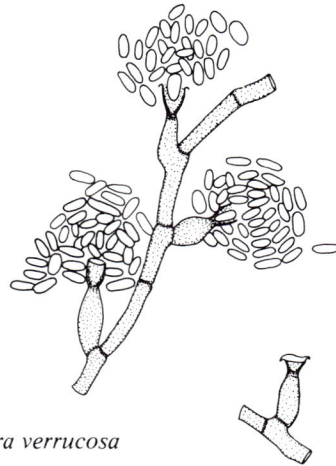

Phialophora verrucosa

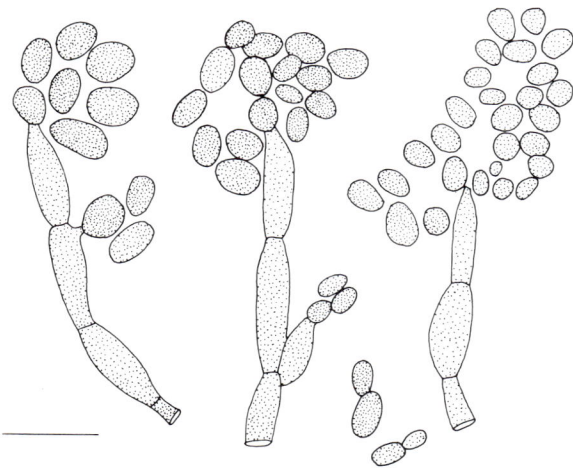

Wangiella dermatitidis

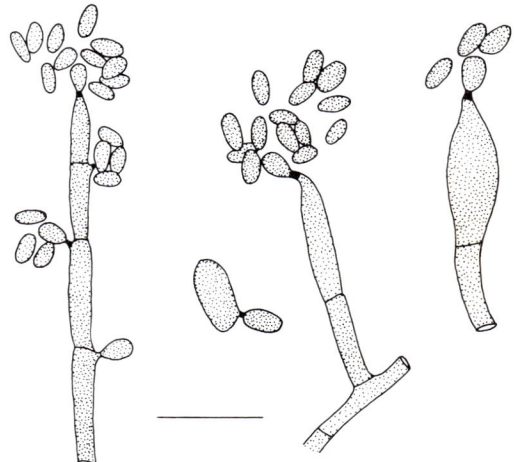

Exophiala jeanselmei

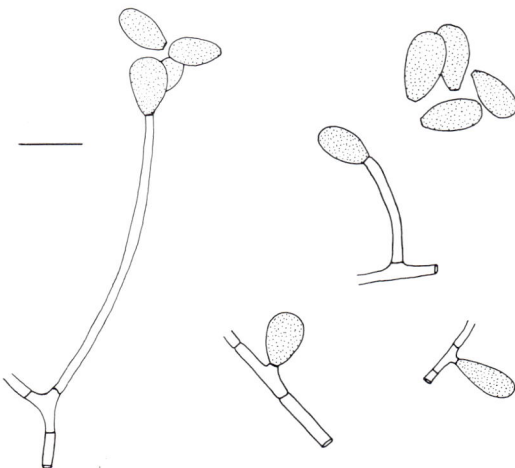

Scedosporium apiospermum

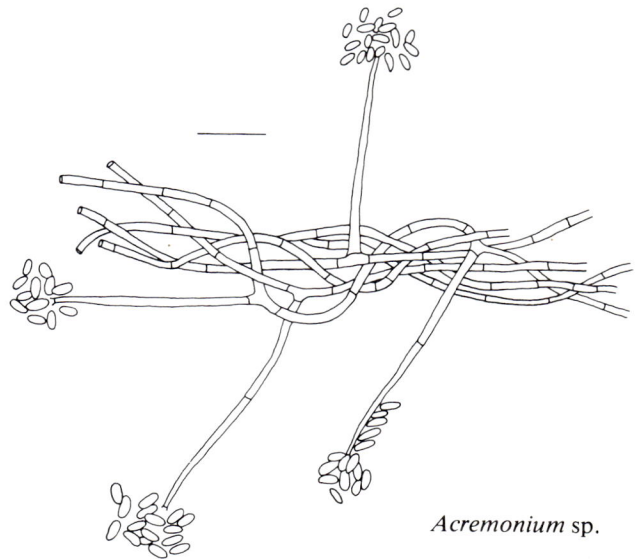

Acremonium sp.

27

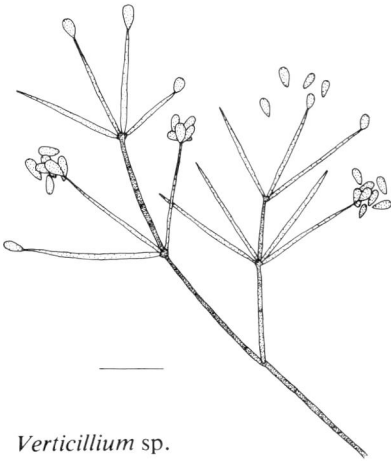

Verticillium sp.

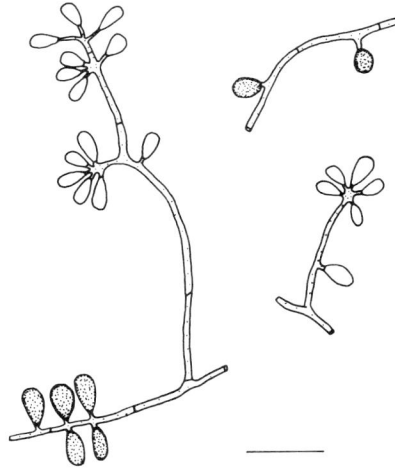

Sporothrix schenckii

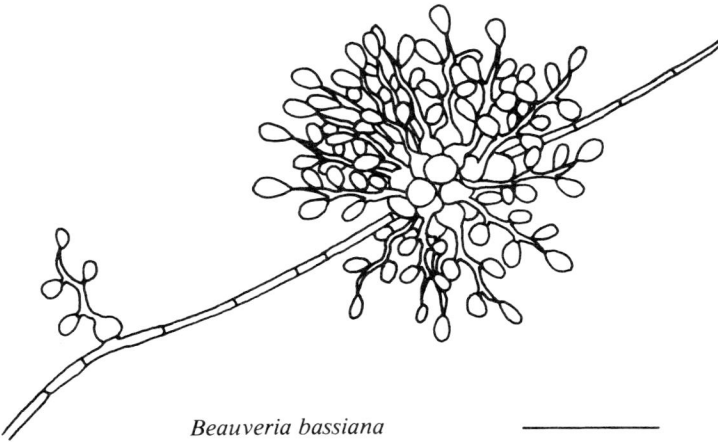

Beauveria bassiana

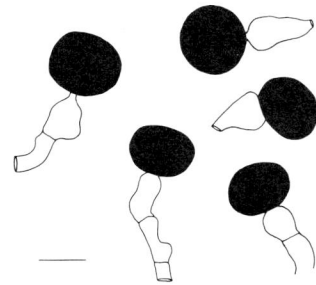

Nigrospora oryzae

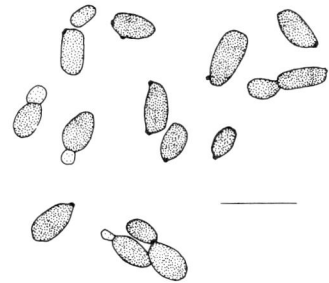

Phaeococcomyces exophialae

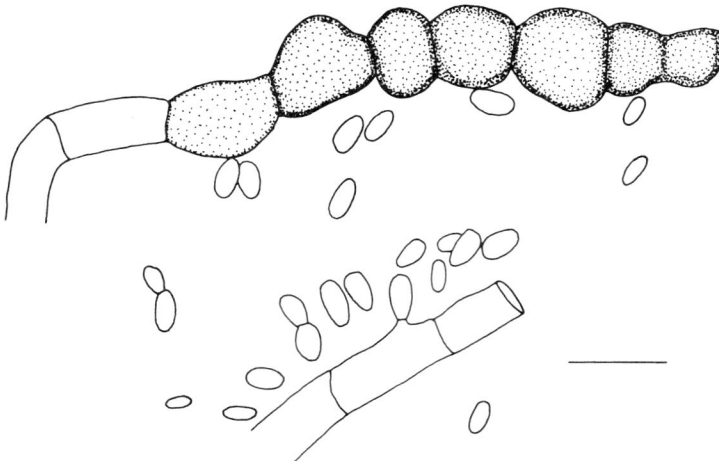

Aureobasidium pullulans

Ustilago violaceum

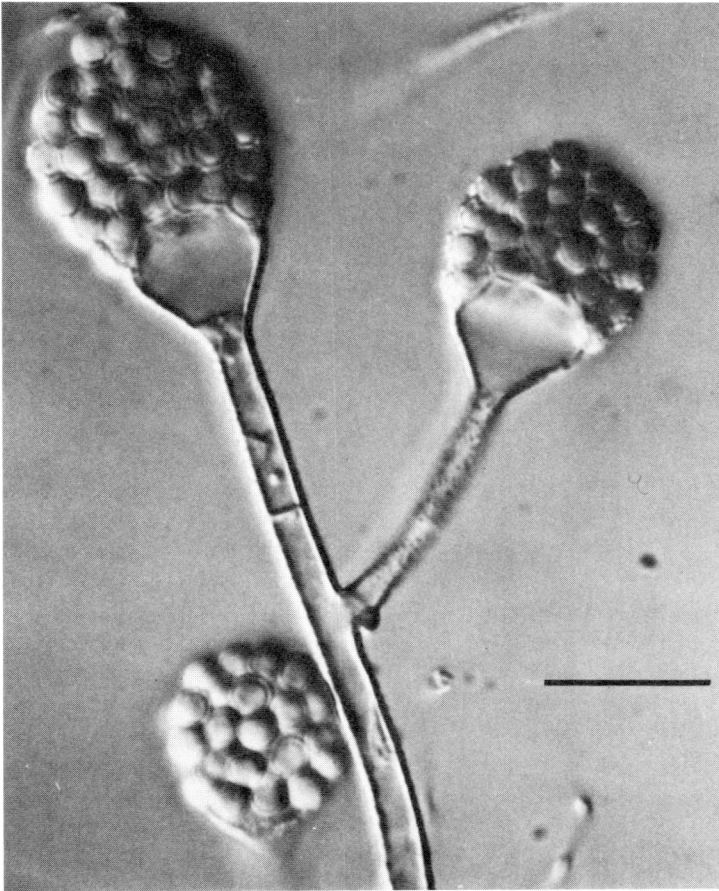

Figure 2. *Absidia* sp. There is a swelling in the sporangiophore where it merges with the columella. Bar is 10 μm.

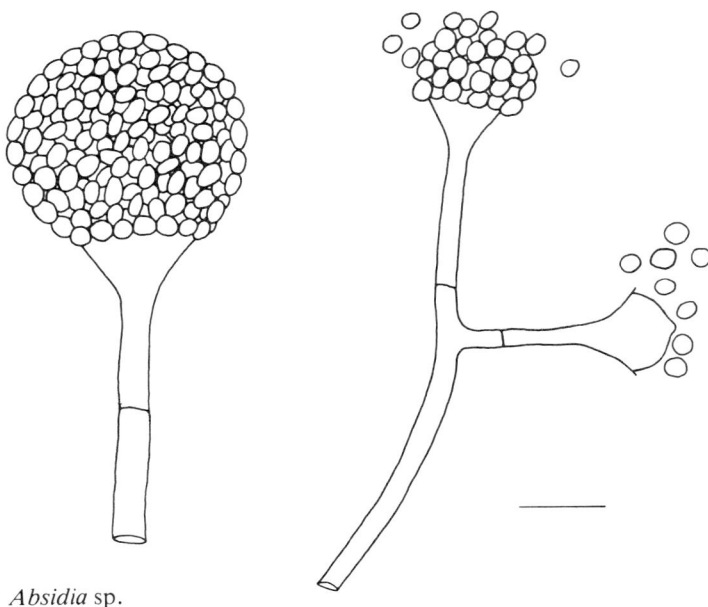

Absidia sp.

Absidia van Tieghem, 1876

Description: Colonies are rapid growing, flat, woolly to cottony, and olive gray in color. Sporangiophores are branched, arising in groups of 2 to 5 at the internodes, often forming arches. A septum is usually present just below the sporangium in the sporangiophore. Sporangiospores are 1-celled, hyaline to light black, globose to ovoid, smooth or rarely echinulate. Rhizoids, when present, are not opposite the sporangiophores.

Salient characteristics: Branched sporangiophores that arise at internodes having an apical swelling that merges with pyriform sporangia are characteristic of the common species of *Absidia*. *Absidia* differ from *Mucor* by having stolons, rhizoids, and branched, apically swollen sporangiophores; from *Rhizopus,* by having rhizoids that do not arise opposite the sporangiophores, and by having apically swollen sporangiophores and sporangiophores that branch. *Absidia* differ from *Rhizomucor* by having pyriform sporangia and apically swollen sporangiophores that merge with the sporangia.

Laboratory precautions: Handle with care, but special precautions are not necessary.

Key reference:
Zycha, Siepmann, and Linnemann, 1969

29

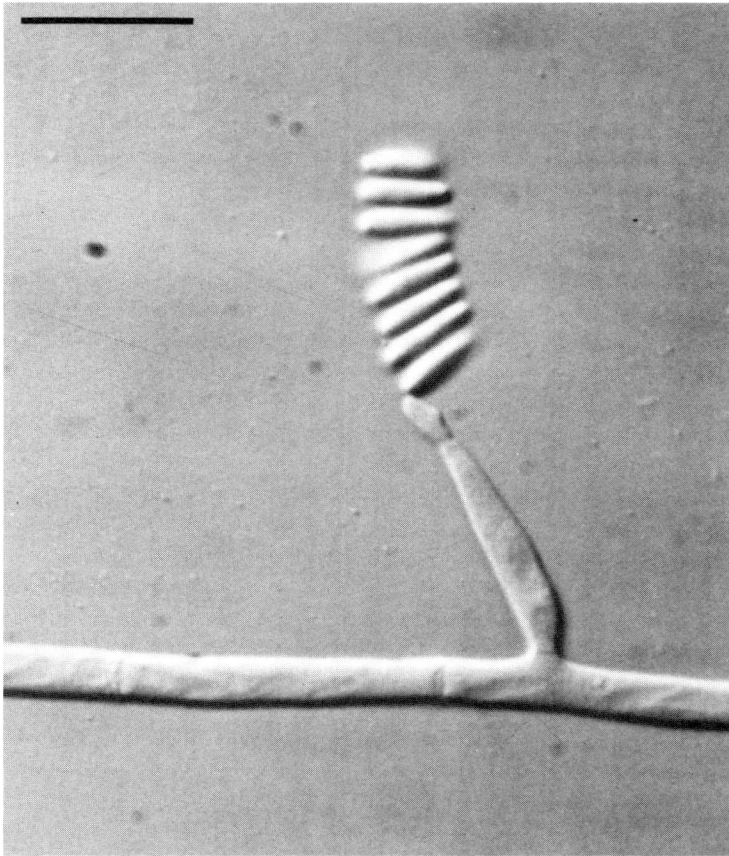

Figure 3. *Acremonium* sp. The 1-celled conidia are accumulating at the apex of the phialide. Bar is 10 μm.

Acremonium Link ex Fries, 1921

Description: Colonies are rapid growing, flat, occasionally raised in the center, velvety to membranelike and glabrous, white, gray, or pink in color. Vegetative hyphae often form hyphal ropes. Phialides are solitary, erect, hyaline and tapering toward their apices, arising directly from the hyphae, the hyphal ropes, or both. Conidia are 1-celled (occasionally 2-celled), hyaline, globose to cylindrical, and accumulating in balls (rarely fragile chains) at the apices of the phialides.

Salient characteristics: *Acremonium* species produce cylindrical, tapering phialides having balls of 1-celled conidia at their apices. *Acremonium* species differ from members of the *Phialophora hoffmannii* complex by having their phialides separated from the hyphae by a septum; and from *Gliomastix,* by having hyaline conidia rather than chains or balls of dark conidia. Species of *Acremonium* are occasionally confused with isolates of *Fusarium, Verticillium,* and *Cylindrocarpon.*

Laboratory precautions: Handle with care, but special precautions are not necessary.

Key reference:
Gams, 1971

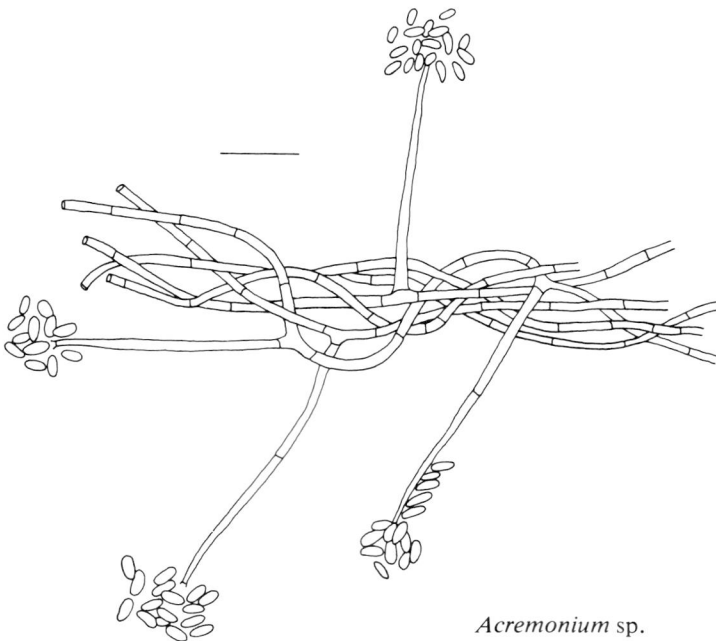

Acremonium sp.

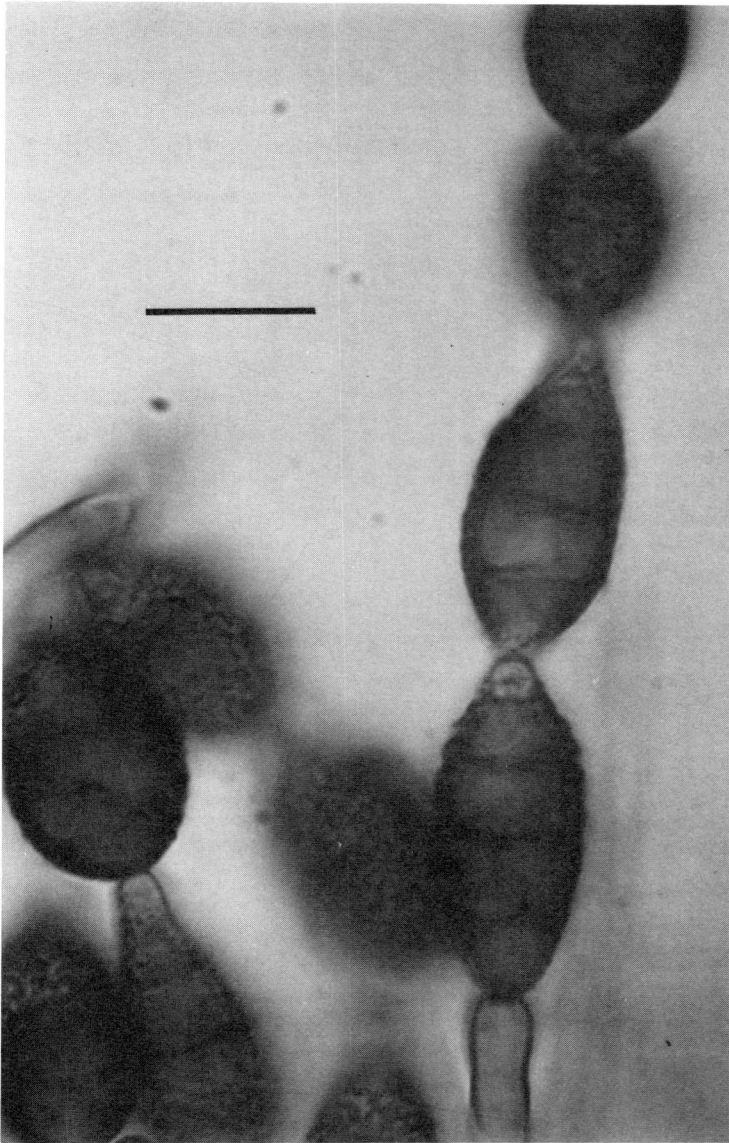

Figure 4. *Alternaria.* sp. The muriform conidia occur in chains. Bar is 10 μm.

Alternaria Nees ex Wallroth, 1833 *nom. cons.*

Description: Colonies are rapid growing, flat, woolly to cottony, gray white to olivaceous in color with a brownish to black reverse coloration. Conidiophores are septate, dark colored, branched or simple, and bear simple or branched, acropetal chains of conidia. Conidia are ovoid to obclavate, darkly pigmented, muriform, smooth or roughened. They are enlarged at their base, tapering toward their apices to form a beak.

Salient characteristics: *Alternaria* species produce chains of darkly pigmented, muriform, obclavate conidia.

Laboratory precautions: Handle with care, but special precautions are not necessary.

Key references:
Joly, 1964,
Simmons, 1967

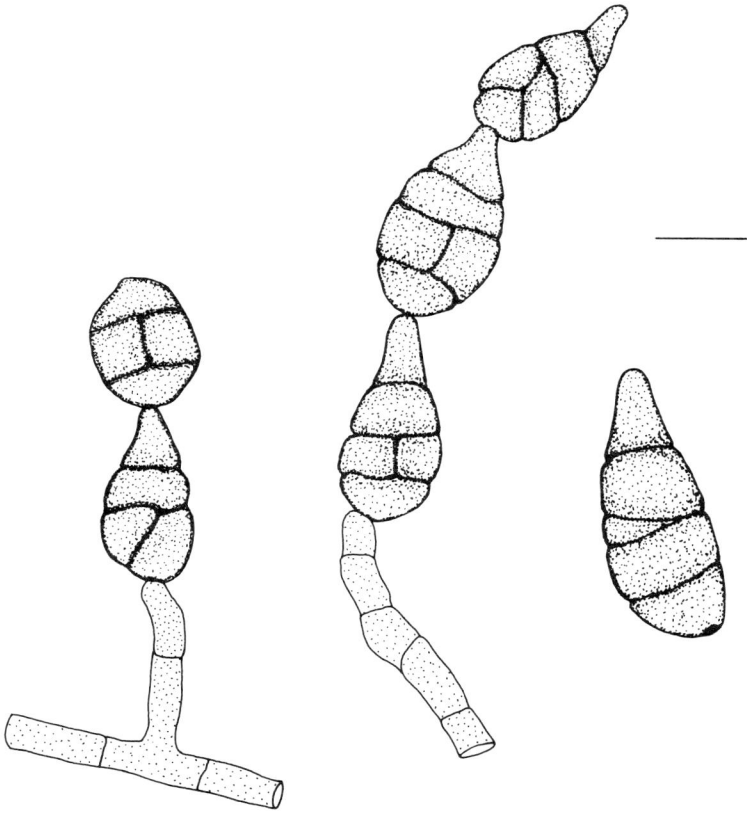

Alternaria sp.

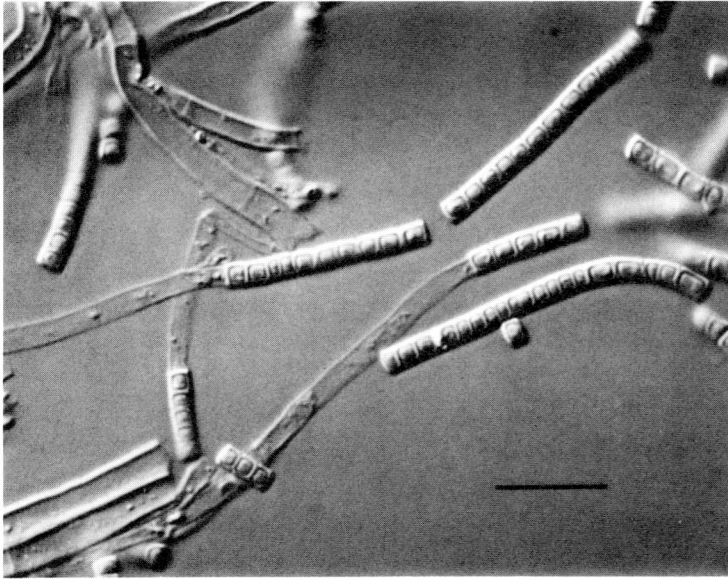

Figure 5. *Arthrographis cuboidea.* The chains of 1-celled arthroconidia develop from conidiophores. Bar is 10 μm.

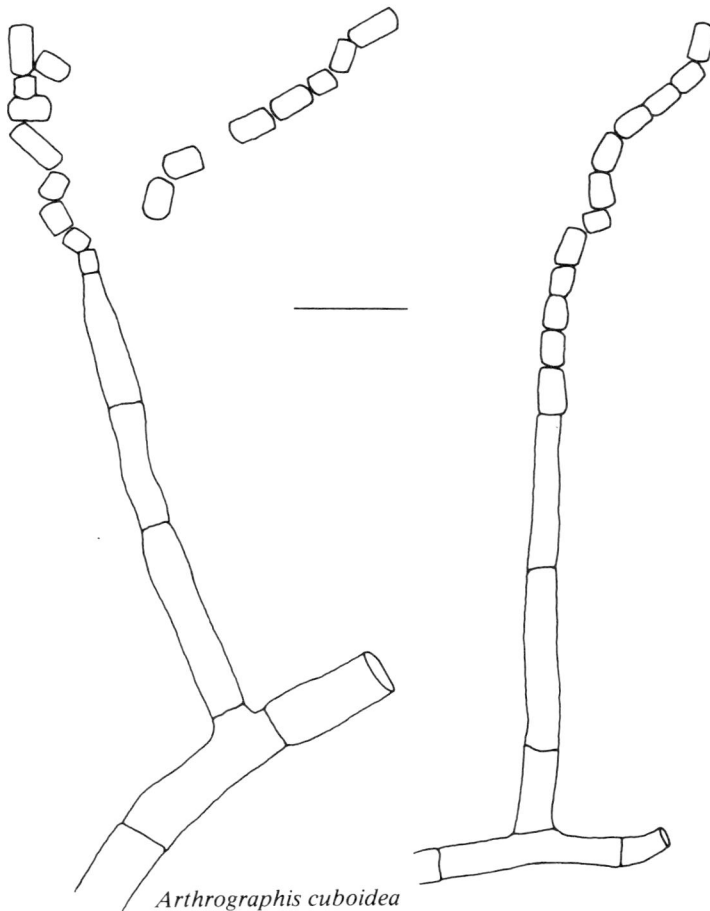

Arthrographis cuboidea

Arthrographis Cochet ex Sigler et Carmichael, 1976

Description: Colonies are moderately fast growing, spreading, smooth, downy, becoming velvety or powdery. Radial ridges or folds may develop. The colonies are creamy white to buff or tan in color with a tan reverse color. Conidiophores are short, branched, and hyaline. Arthroconidia are of two types: conidia of the first type arise from conidiophores and are 1-celled, cylindrical, hyaline, smooth, and in chains; conidia of the second type arise from undifferentiated hyphae and are longer and narrower. The conidia are released by fission through double septa.

Salient characteristics: Branching chains of hyaline arthroconidia arising upon short, hyaline conidiophores distinguish the more common species of *Arthrographis*. *Arthrographis* differs from *Geotrichum* by having distinct conidiophores; and from *Oidiodendron* by having hyaline conidiophores rather than pigmented ones.

Laboratory precautions: Handle with care, but special precautions are not necessary.

Key reference:
Sigler and Carmichael, 1976

33

Aspergillus Micheli ex Link, 1821

Description: Colonies are rapid growing, woolly to cottony, typically in some shade of green, brown, or black. Conidiophores are erect, with a swollen vesicle at their apices, often with a foot cell where they merge with the hyphae, simple, hyaline to brown in color. Phialides are flask-shaped, arising from the vesicles (uniseriate), from metulae on the vesicles (biseriate), or from both. Conidia are 1-celled, globose to subglobose, smooth or echinulate, occurring in chains.

Salient characteristics: *Aspergillus* species form chains of 1-celled conidia that arise from phialides on conidiophores having apical vesicles and, usually, foot cells.

Laboratory precautions: Handle with care, but special precautions are not necessary.

Key references:
Raper and Fennell, 1965
Samson, 1979

Key to the More Common Species of *Aspergillus*

1. Heads are uniseriate only... 2
1.′ Heads are biseriate, or are both uniseriate and biseriate 3
    2. Vesicles clavate with phialides covering the entire surface; conidia elliptic, smooth, with thick walls...*A. clavatus*
    2.′ Vesicles hemispherical with phialides covering the upper half; conidia globose to subglobose, echinulate or delicately roughened, 2–3.5 μm in diameter; conidial heads forming compact columns; good growth at 45°C...*A. fumigatus*
3. Heads are biseriate only... 4
3.′ Uniseriate or biseriate conditions seen in the same isolate or on a single vesicle; conidiophores roughened; conidial heads yellow green, radiating, splitting into several poorly defined columns... *A. flavus*
    4. Conidial heads black; vesicles globose, with large metulae and small phialides; conidial heads radiating around vesicle... *A. niger*
    4.′ Conidial heads green, cinnamon, buff to orange brown...5
5. Conidial heads green; conidiophores sinuous, cinnamon brown, with a hemispherical vesicle; conidial heads as short columns; cleistothecia and hulle cells typically present... *A. nidulans*
5.′ Conidial heads cinnamon, buff to orange brown; conidiophores hyaline with a hemispherical vesicle; conidial heads as long, compact columns .... *A. terreus*

Aspergillus clavatus Desmazières, 1834

Description: Colonies on Czapek-Dox solution agar are rapid growing, thin to cottony, flat to slightly furrowed and blue-green with an uncolored reverse that becomes brown with age. Conidiophores are evenly distributed or are in zones, thin walled, smooth, hyaline, with a large clavate vesicle. Phialides are flask shaped, uniseriate, covering the entire vesicle. Conidia are 1-celled, elliptical, smooth, thick walled, 2.5–3.5 μm. Conidial heads split into 2, 3, or more compact, divergent columns.

Salient characteristics: *Aspergillus clavatus* forms large, clavate vesicles with a uniseriate arrangement of phialides and a few divergent columns of conidia.

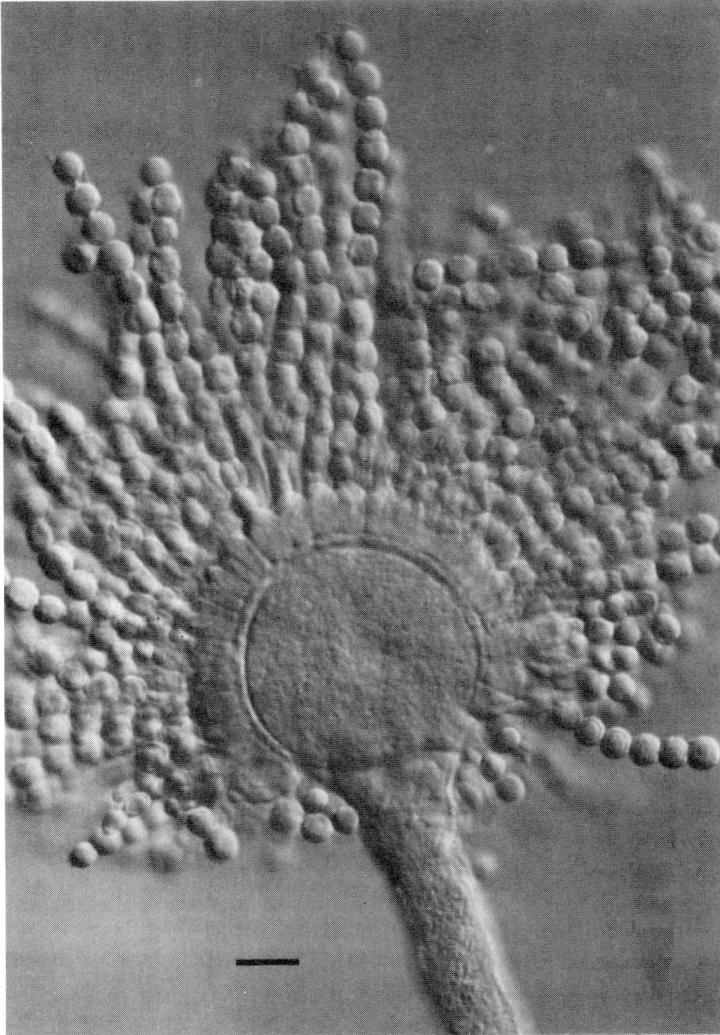

Aspergillus flavus
Link ex Link, 1824

Description: Colonies on Czapek-Dox solution agar are slow to rapid growing, woolly to cottony, occasionally radially furrowed, initially yellow, quickly becoming bright to dark yellow green or jade green, with an uncolored to pinkish drab reverse color. Sclerotia are often present. Conidiophores are thick walled, hyaline, coarsely roughened with a subglobose or globose vesicle. Phialides are flask shaped, either uniseriate or biseriate, covering the majority of the vesicle. Conidia are 1-celled, globose to subglobose, echinulate, 3–6 μm. Conidial heads radiate, typically splitting into several poorly formed columns.

Salient characteristics: *Aspergillus flavus* is characterized by pale to intense yellow green colonies, coarsely roughened conidiophores, conidia arising from phialides that are either uniseriate or biseriate in arrangement, and radiating conidial heads that split into loosely organized columns.

Figure 6. *Aspergillus flavus.* The conidial head radiates about the vesicle with the phialides in a uniseriate arrangement. Bar is 10 μm.

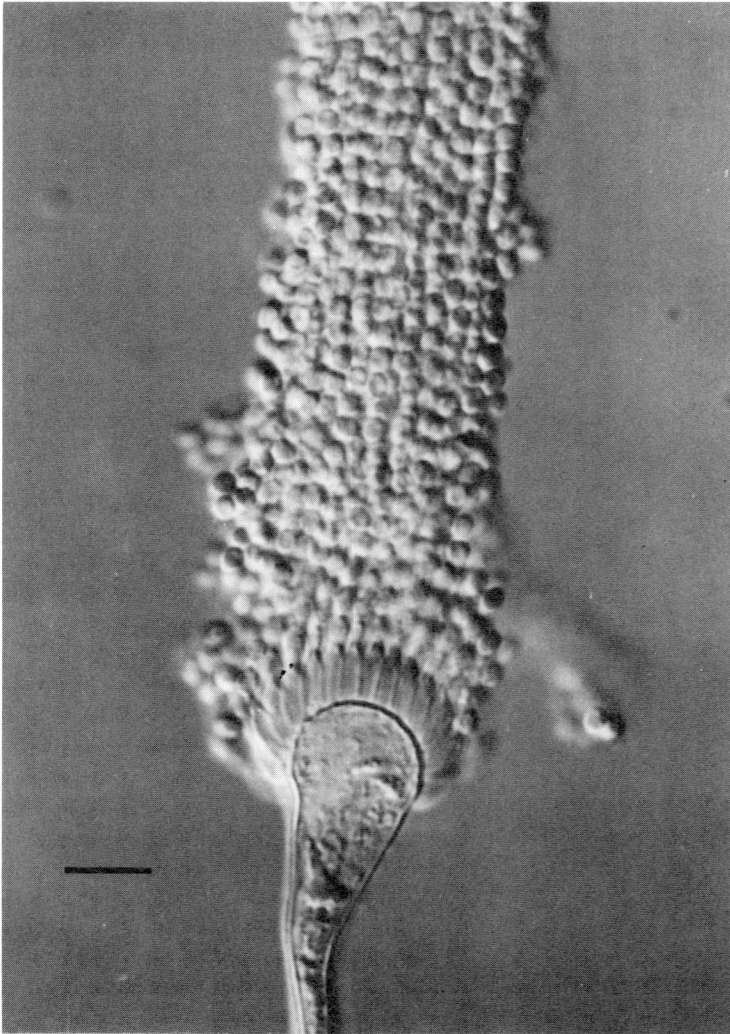

Figure 7. *Aspergillus fumigatus.* The conidial head is columnar with the phialides in a uniseriate arrangement. Bar is 10 μm.

Aspergillus fumigatus
Fresenius, 1863

Description: Colonies on Czapek-Dox solution agar are rapid growing, velvety to cottony, at first white, becoming green to dark gray with an uncolored, or yellow, green, or dark brown reverse. Colonies grow at 45°C. Conidiophores are thin walled, smooth, green, ending in a hemispherical vesicle. Phialides are flask shaped, uniseriate, covering the upper half of the vesicle. Conidia are 1-celled, globose to sub-globose, echinulate, thin walled, 2–3.5 μm. Conidial heads form a single, compact column.

Salient characteristics: *Aspergillus fumigatus* produces flask-shaped vesicles bearing phialides in a uniseriate arrangement, a single, compact column, conidia that are 2–3.5μm in diameter, and colonies that grow at 45°C. *Aspergillus fumigatus* differs from the *A. fischeri* series by the fact that it lacks cleistothecia

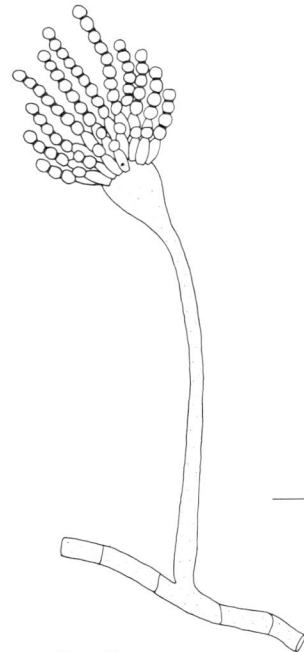

Aspergillus fumigatus

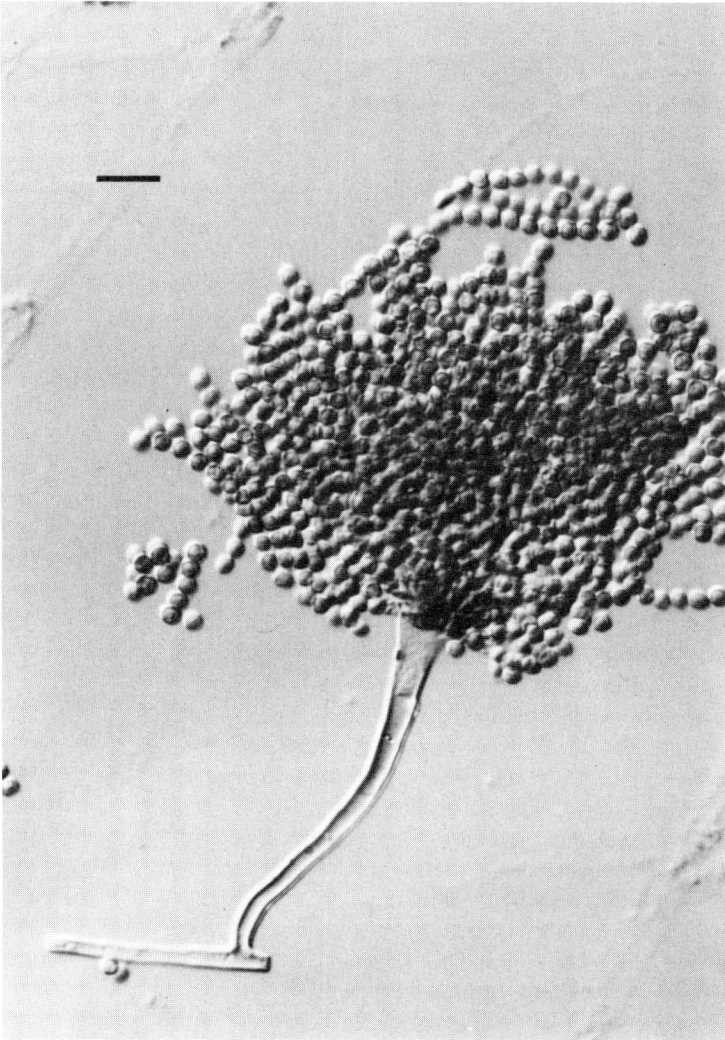

Aspergillus nidulans
(Eidam) Winter, 1884

Description: Colonies on Czapek-Dox solution agar are rapid growing, velvety to cottony, dark green, with a purplish red reverse color. Conidiophores are sinuous, smooth, cinnamon brown, with a hemispherical vesicle. Phialides are flask shaped, biseriate, covering the upper half of the vesicle. Conidia are 1-celled, globose, rough walled, 3–3.5 μm. Conidial heads are short columns. Cleistothecia are typically abundant.

Salient characteristics: *Aspergillus nidulans* forms hemispherical vesicles bearing a biseriate arrangement of phialides and green conidial heads occurring as short columns.

Figure 8. *Aspergillus nidulans.* The conidial head covers the upper half of the vesicle. Bar is 10 μm.

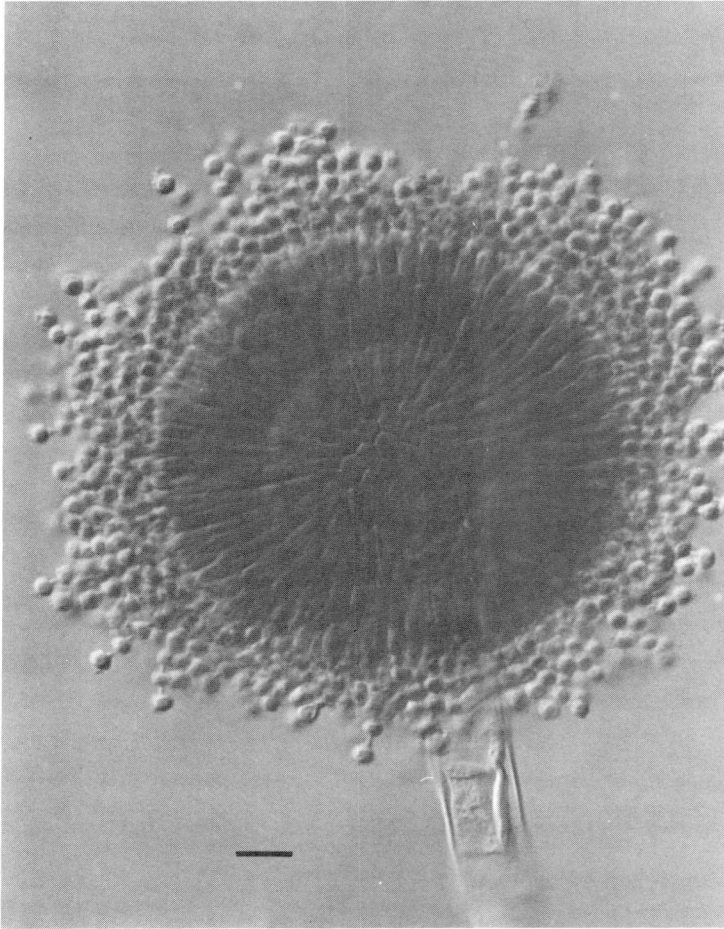

Aspergillus niger
van Tieghem, 1867

Description: Colonies on Czapek-Dox solution agar are rapid growing, compact, granular, at first white, becoming carbonaceous black, often with a pale yellow reverse color. Conidiophores are smooth, thick walled, colorless or brown, having a large, globose vesicle. Phialides are flask shaped, biseriate, covering the entire vesicle. Metulae are larger than the small phialides. Conidia are 1-celled, globose, thick walled, roughened or echinulate, with ridges, and are 4–5 μm in diameter. Conidial heads radiate, often splitting into two or more loose columns with age.

Salient characteristics: *Aspergillus niger* forms black colonies that have a pale yellow reverse color, and conidiophores with globose vesicles having biseriately arranged phialides, and radiate conidial heads.

Figure 9. *Aspergillus niger.* The black conidial head radiates about the vesicle with the phialides in a biseriate arrangement. Bar is 10 μm.

Aspergillus terreus Thom, 1918

Description: Colonies on Czapek-Dox solution agar are rapid growing, velvety to cottony, flat or with shallow radial furrows, cinnamon, buff to brown, with a dull yellow to brown reverse color. Some isolates produce an amber-colored exudate. Conidiophores are smooth, hyaline, and have a dome-like vesicle. Phialides are flask shaped, biseriate, covering most of the vesicle surface. Conidia are 1-celled, globose to slightly elliptic, smooth, 1.8–2.4 μm. Conidial heads are long, forming compact columns.

Salient characteristics: *Aspergillus terreus* forms cinnamon or buff to orange brown colonies, phialides in a biseriate arrangement, and long, compact columnar conidial heads.

Figure 10. *Aspergillus terreus.* The conidial head is beginning to form a column with the phialides in a biseriate arrangement. Bar is 10 μm.

40

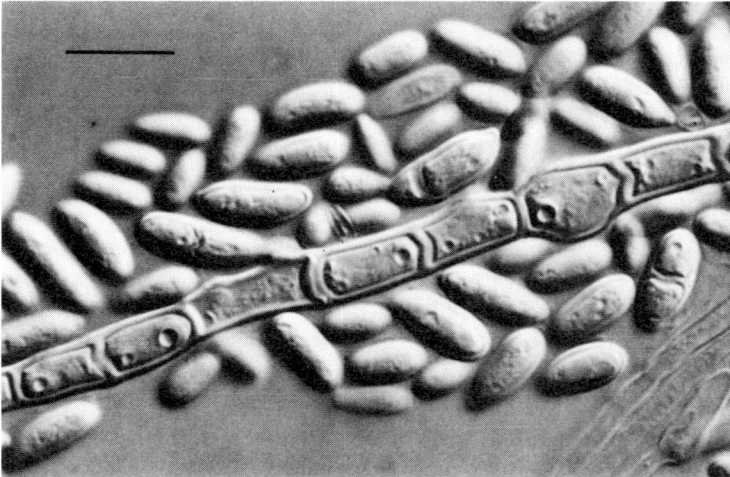

Figure 11. *Aureobasidium pullulans.* The 1-celled conidia are arising from the hypha. Bar is 10 μm.

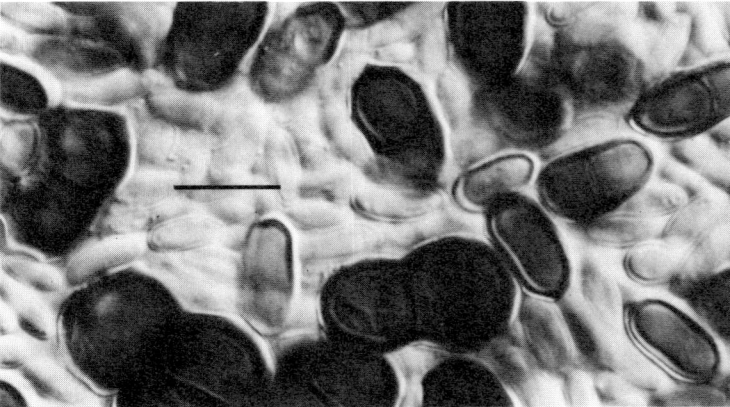

Figure 12. *Aureobasidium pullulans.* Dark-colored intercalary conidia are often formed. Bar is 10 μm.

Aureobasidium
Viala et Boyer, 1891

Description: Colonies are moderately rapid growing, flat, smooth, often moist and mucoid to pasty in appearance, yellow or pink, turning light brown, or more frequently black with age. Some aerial hyphae may be present. Distinct conidiophores are absent. Conidia are 1-celled, hyaline, oval to cylindrical, arising in clusters or along the hyphae. Conidia often form secondary blastoconidia. One- to 2-celled, darkly pigmented arthroconidia are usually present.

Salient characteristics: Isolates of *Aureobasidium* produce 1-celled hyaline conidia (solitary or in clusters) along the hyphae, and 1- or 2-celled, thick-walled, dematiaceous arthroconidia. *Aureobasidium* differs from *Phaeococcomyces* by having hyaline-budding conidia. Both of these genera form black, pasty colonies.

Laboratory precautions: Handle with care, but special precautions are not necessary.

Key reference:
Hermanides-Nijhof, 1977

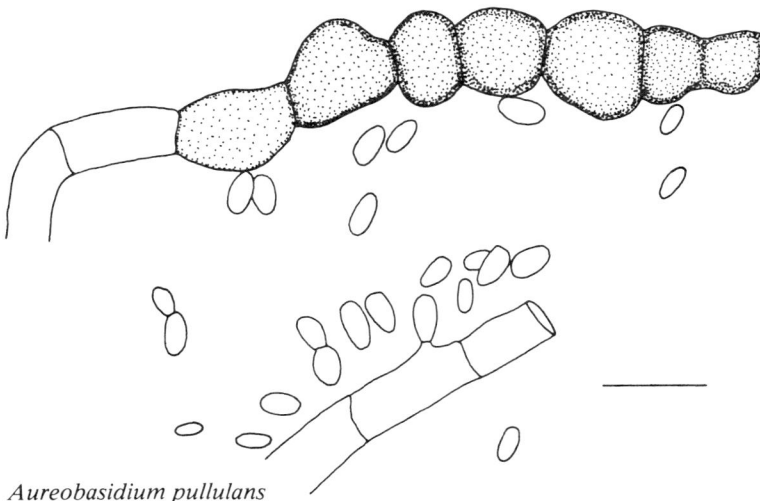

Aureobasidium pullulans

41

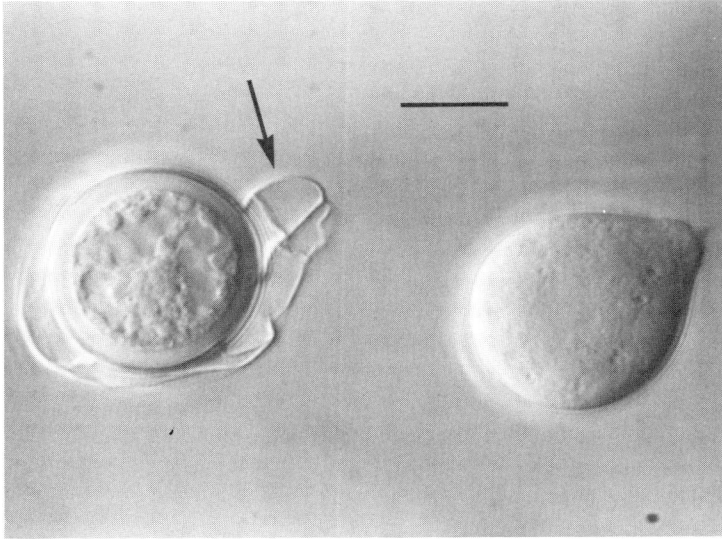

Figure 13. *Basidiobolus ranarum.* Beaks (arrow) are associated with the zygospore. Bar is 10 μm.

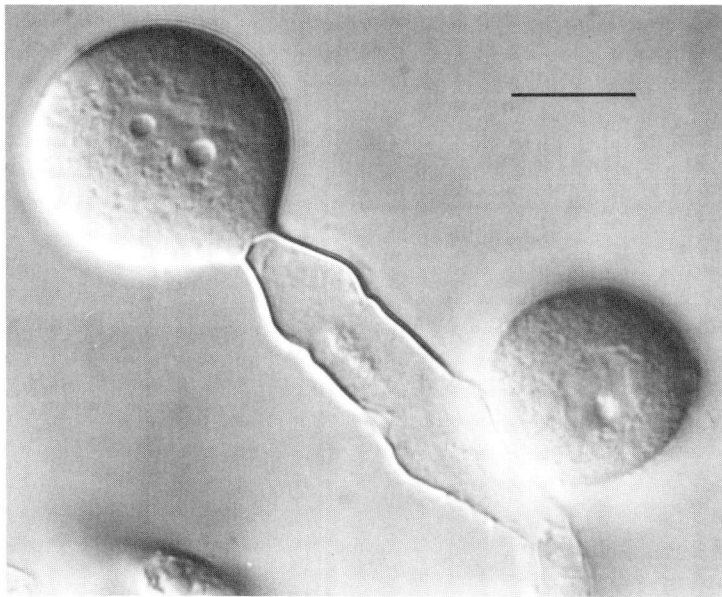

Figure 14. *Basidiobolus ranarum.* Some spores are forcibly discharged from a sporangiophore that is slightly swollen. Bar is 10 μm.

Basidiobolus Eidam, 1886

Description: Colonies are rapid growing, initially flat, becoming heaped and folded, waxy in texture with some short aerial hyphae developing with age, and yellowish white in color. Zygospores are formed within the larger of the two gametangia, with two lateral protuberances (beaks) typically present; zygospores are thick walled, smooth or with undulating outer cell walls. Primary sporangiola are unispored, globose to pyriform, and forcibly discharged from the apex of a sporangiophore that has a subterminal swelling. Upon germination, the primary sporangiola may form

1. a secondary sporangiolum like the primary one;
2. several sporangiola, called microspores, that develop upon pedicels around the primary sporangiolum;
3. a sporangiolum containing several sporangiospores;
4. a solitary, elongate, obclavate sporangiolum with an adhesive tip.

Salient characteristics: *Basidiobolus* is characterized by thick-walled zygospores having two beaklike protuberances arising laterally from the gametangia and by forcibly discharged 1-celled primary sporangiola. *Basidiobolus* differs from *Conidiobolus* by having beaks associated with the gametangia.

Laboratory precautions: Handle with care, but special precautions are not necessary.

Key references:
Fuller, 1978
O'Donnell, 1979

42

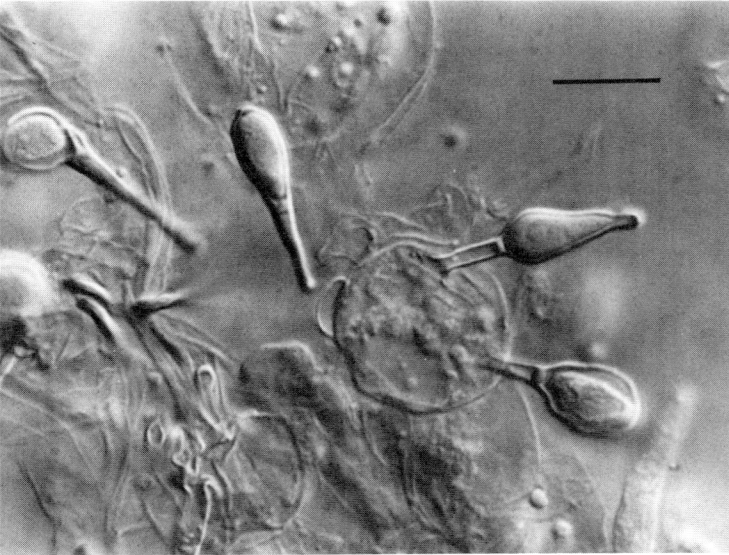

Figure 15. *Basidiobolus microsporus.* Microspores are produced by this species. Bar is 10 μm.

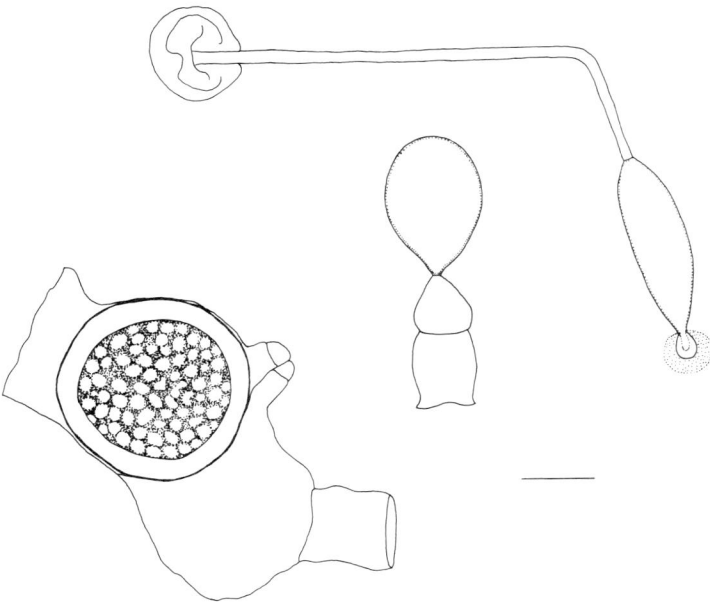

Basidiobolus ranarum

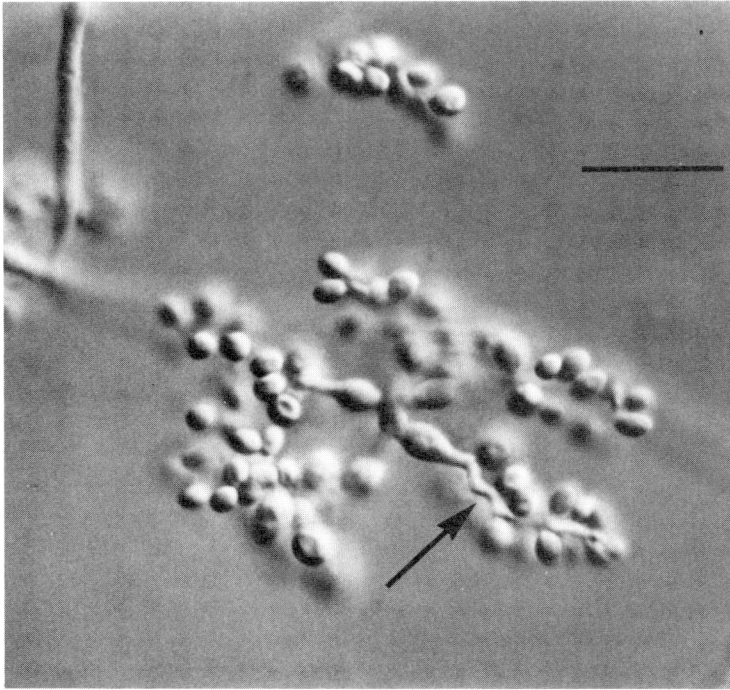

Figure 16. *Beauveria bassiana.* The 1-celled conidia develop along a geniculate rachis (arrow). Bar is 10 μm.

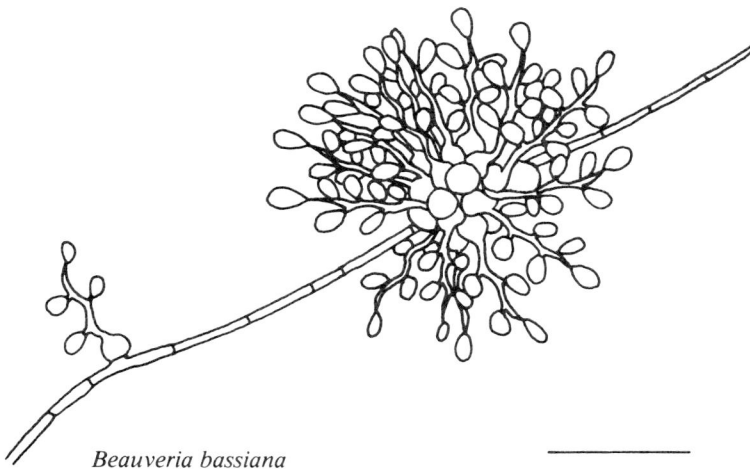

Beauveria bassiana

Beauveria Vuillemin, 1912

Description: Colonies are moderately rapid growing, spreading, woolly, powdery, or mealy in texture, white to yellowish white or occasionally pinkish in color. Conidiogenous cells are hyaline, flask shaped with a long zigzag-appearing rachis bearing lateral conidia. Conidia are hyaline, 1-celled, and globose to ovoid. Clusters of conidiogenous cells appear as small powdery balls in the aerial hyphae when viewed through a dissecting microscope.

Salient charactersitics: Clusters of flask-shaped cells, each with a long zigzag-appearing rachis bearing 1-celled conidia laterally are distinctive for members of the genus *Beauveria.* The clusters of swollen conidiogenous cells with a zigzagging rachis distinguish *Beauveria* from similar fungi, such as species of *Tritirachium.*

Laboratory precautions: Handle with care, but special precautions are not necessary.

Key reference:
de Hoog, 1972

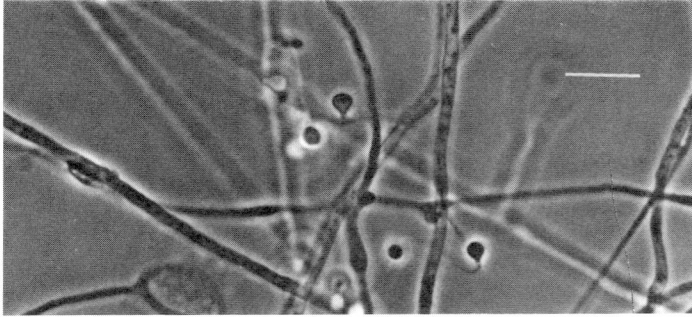

Figure 17. *Blastomyces dermatitidis.* The globose 1-celled conidia arise from short conidiophores, 25°C. Bar is 10 μm. (Reproduced by permission of Academic Press from M. McGinnis. *Laboratory Handbook of Medical Mycology.* 1980.)

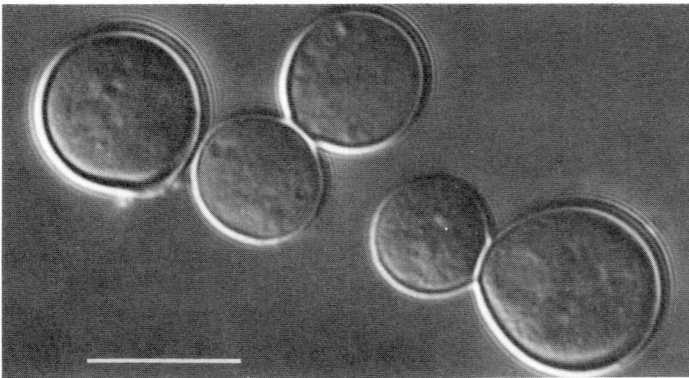

Figure 18. *Blastomyces dermatitidis.* The large yeast with broadly attached blastoconidia developed at 37°C. Bar is 10 μm. (Reproduced with permission of Academic Press from M. McGinnis. *Laboratory Handbook of Medical Mycology.* 1980.)

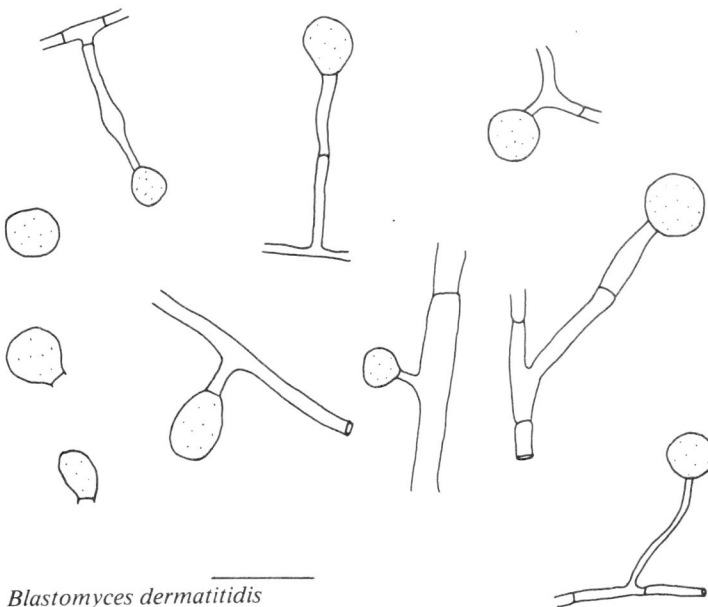

Blastomyces dermatitidis

Blastomyces Gilchrist et Stokes, 1898

Description: Colonies are slow to moderately rapid growing, membranous to woolly in texture, and white to tan with a tan to brownish reverse color. At 25°C, conidiophores are hyphalike, arising at right angles to the vegetative hyphae. Conidia are hyaline, 1-celled, solitary, globose to subglobose. At 37°C, a yeast form (with globose, thick-walled blastoconidia that are attached to the parent cell by a broad base) is formed.

Salient characteristics: *Blastomyces dermatitidis,* the only species in the genus, is recognized by its distinctive globose to subglobose conidia that arise on short conidiophores, as well as by its conversion to a yeast form at 37°C. The globose blastoconidia are attached to the parent cells by a broad base. *Blastomyces* is readily distinguished from *Chrysosporium* by the fact that it is dimorphic. A specific and sensitive exoantigen serologic procedure is available for the rapid and specific identification of cultures of *B. dermatitidis.*

Laboratory precautions: Handle with care in a biological safety cabinet. This fungus is potentially dangerous.

Key references:
Carmichael, 1962
McGinnis, 1980
van Oorschot, 1980

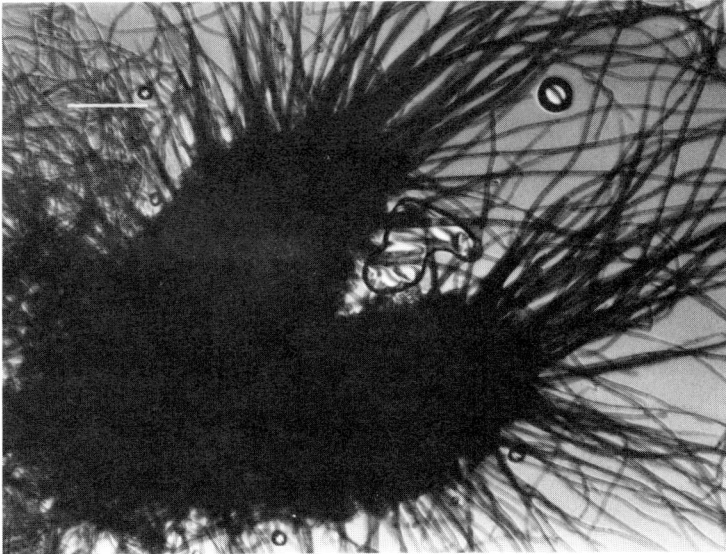

Figure 19. *Chaetomium* sp. The perithecia have long hairlike setae. Bar is 10 μm.

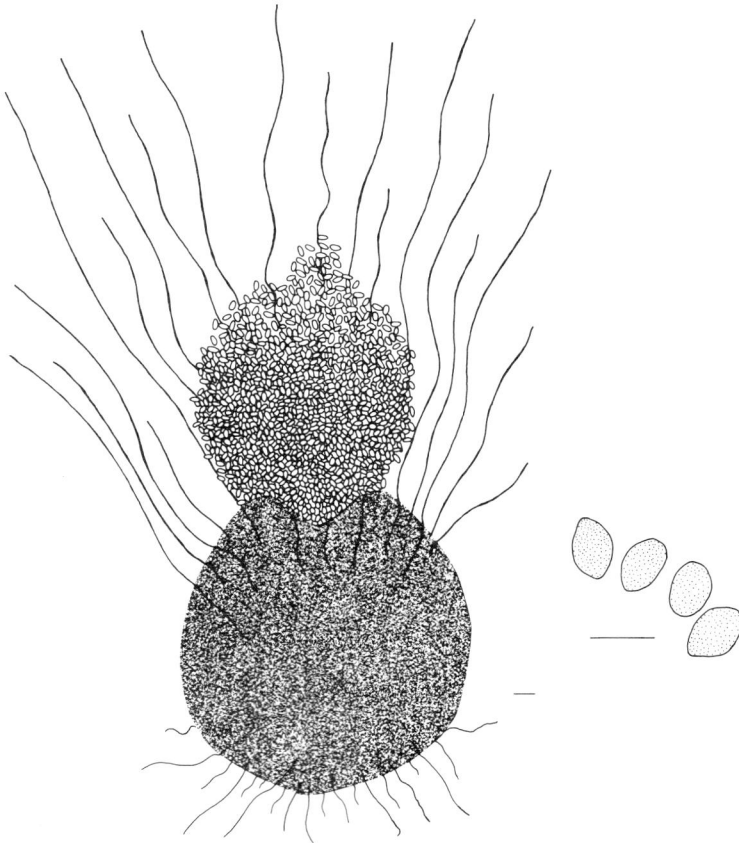

Chaetomium sp.

Chaetomium
Kunze ex Fries, 1829

Description: Colonies are rapid growing, cottony, at first white, becoming gray, grayish olive to dark olive green with olive brown to black reverse color. Perithecia are superficial, dark brown to black, globose to flask shaped, with long, hairlike, brown to black setae. The perithecia have ostioles. The fruiting body is brittle and fragile. Asci are clavate to cylindrical, rapidly dissolving to release their eight ascospores. Ascospores are 1-celled, olive brown, and lemon shaped.

Salient characteristics: Isolates of *Chaetomium* are recognized by their dark, brittle perithecia having long, hairlike appendages and by their 1-celled, olive brown, lemon-shaped ascospores.

Laboratory precautions: Handle with care, but special precautions are not necessary.

Key reference:
Ames, 1961

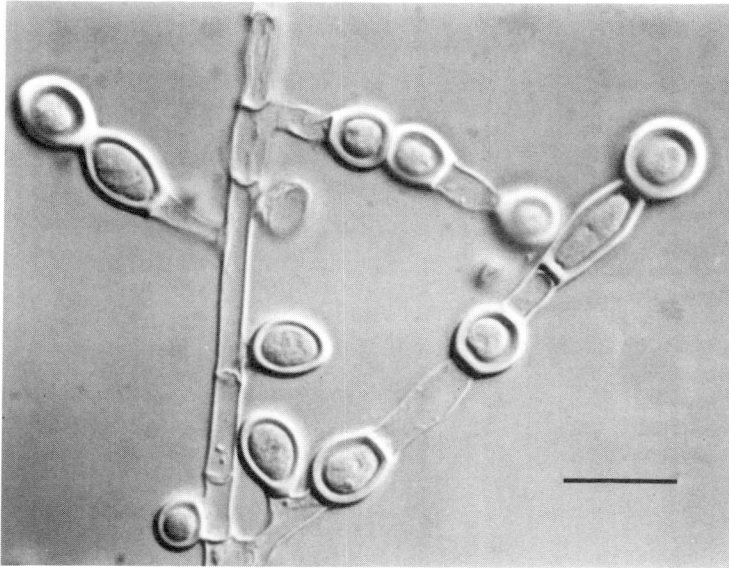

Figure 20. *Chrysosporium xerophilum.* The conidia develop intercalary and upon short conidiophores. Bar is 10 μm.

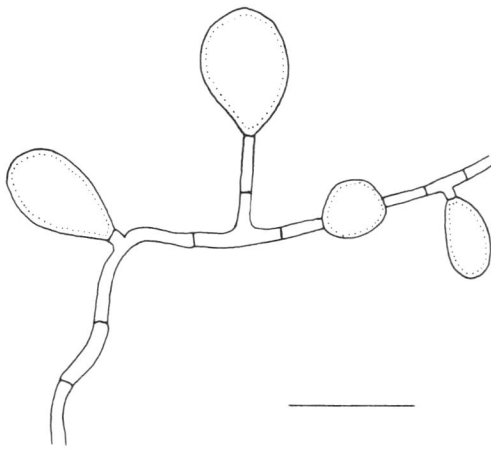

Chrysosporium sp.

Chrysosporium Corda, 1833

Description: Colonies are moderately rapid growing, granular, woolly, or cottony, dry, flat, raised, or folded, white cream, yellow, or tan to pale brown, with a white to brown reverse color. Distinct conidiophores are absent. Conidia are hyaline, 1-celled, smooth or rough walled, and broader than the vegetative hyphae; they occur terminally on pedicels, along the sides of the hyphae, or both. Conidia have a broad base and an annular frill that is formed when the conidia separate from their conidiophores. Random arthroconidia that are larger in diameter than their parent hyphae typically are present in abundant numbers.

Salient characteristics: *Chrysosporium* species form hyaline, 1-celled, lateral conidia in conjunction with numerous 1-celled random arthroconidia. *Chrysosporium* differs from *Blastomyces* by not being dimorphic; from *Microsporum* and *Trichophyton* by lacking macroconidia; from *Geomyces* by lacking branched, fertile hyphae on erect conidiophores; and from *Sepedonium* by having hyaline conidia. The principal species of interest is *C. parvum,* which includes the two fungi previously called *Emmonsia crescens* and *E. parva. Chrysosporium parvum* forms enlarged cells (adiaspores) at 37–40°C.

Laboratory precautions: Handle with care, but special precautions are not necessary.

Key references:
Carmichael, 1962
McGinnis, 1980
van Oorschot, 1980

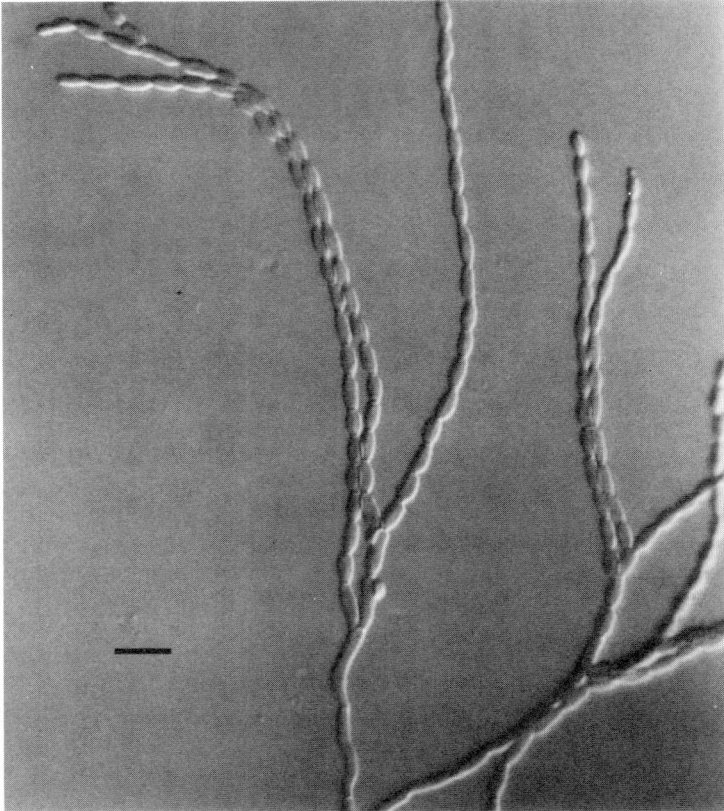

Figure 21. *Cladosporium bantianum.* The chains of 1-celled dematiaceous conidia are sparsely branched. Bar is 10 μm.

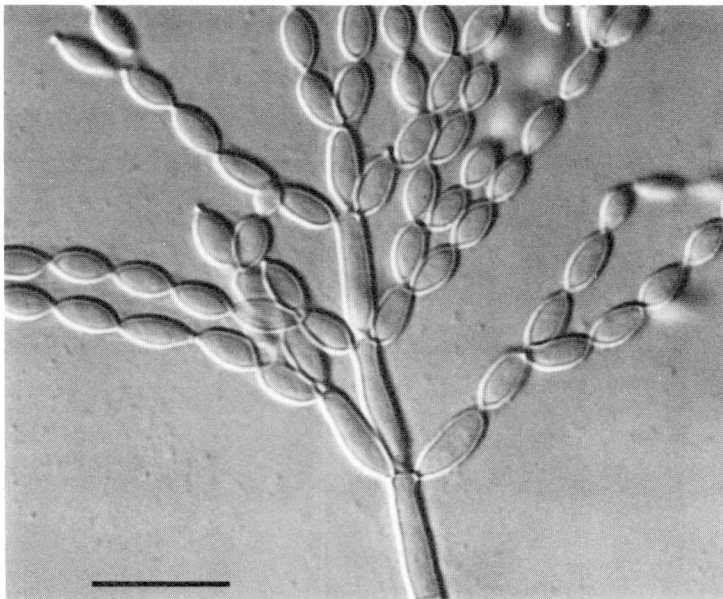

Figure 22. *Cladosporium carrionii.* The 1-celled conidia occur in branching chains. Bar is 10 μm.

Cladosporium
Link ex Gray, 1821

Description: Colonies are moderately rapid growing, spreading, velvety to woolly, grayish green to olivaceous green with a darkly colored reverse. Conidiophores are erect, septate, and pigmented. Conidia are 1- to 4-celled, with a dark hilum, pale brown to dark brown, smooth or occasionally echinulate, and occur in branching chains that readily disarticulate. Some conidia are shield shaped.

Salient characteristics: Isolates of *Cladosporium* produce branching, fragile chains of dematiaceous blastoconidia with dark hila, some of which are shield shaped. The conidia arise from an erect dematiaceous conidiophore. *Cladosporium bantianum* grows at 42–43°C and has conidia that are approximately 6.4 μm in length. *Cladosporium carrionii* does not grow at temperatures beyond 35–36°C and has conidia that are approximately 4.8–5.2 μm long.

Laboratory precautions: *Cladosporium bantianum* is an extremely dangerous fungus that should be worked with only in a biological safety cabinet. *Cladosporium carrionii* also should be handled with care in a biological safety cabinet.

Key references:
de Vries, 1952
McGinnis, 1980
McGinnis and Borelli, 1981

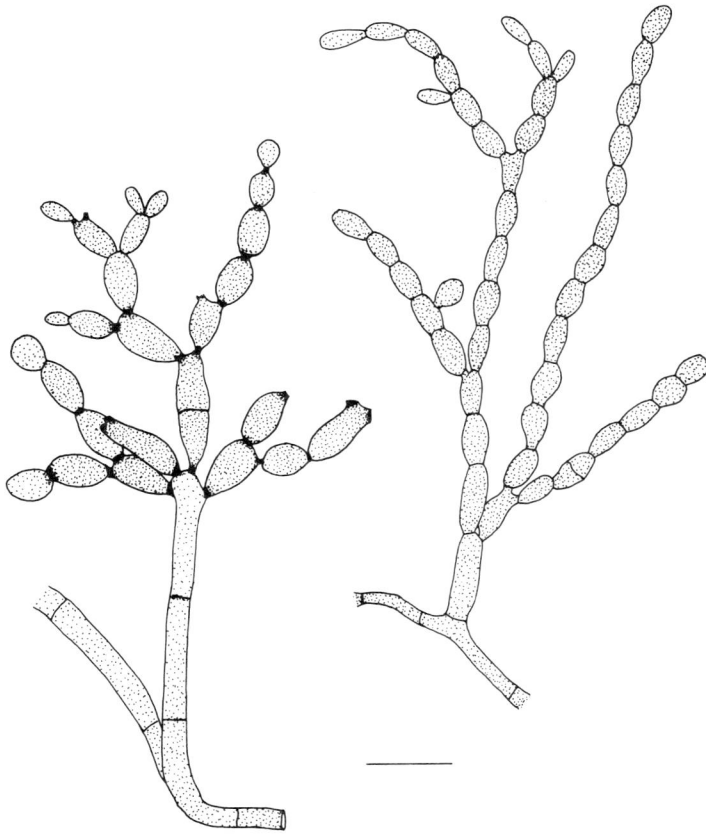

Cladosporium sp.

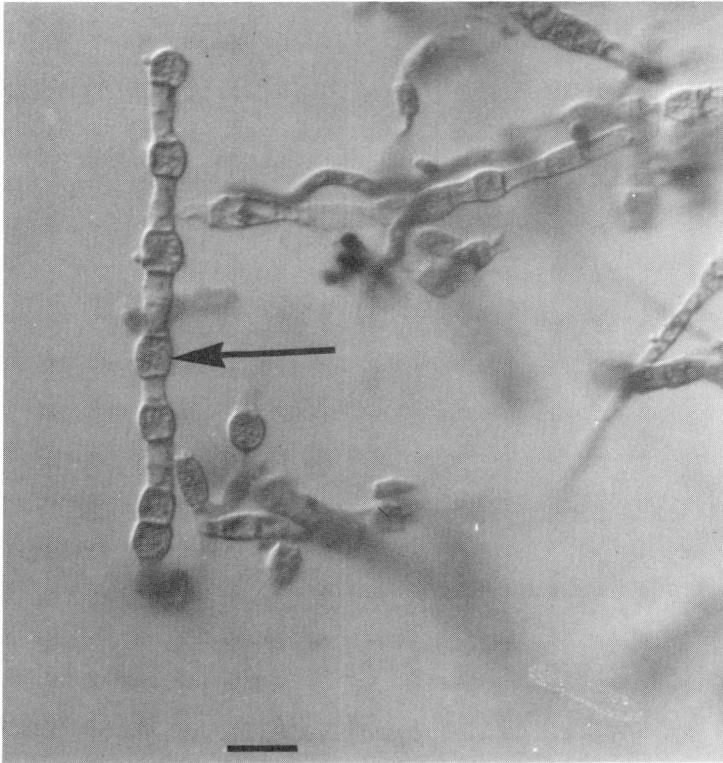

Figure 23. *Coccidioides immitis.* The arthroconidia (arrow) are separated from each other by disjunctor cells, 25°C. Bar is 10 μm. (Reproduced by permission of Academic Press from M. McGinnis. *Laboratory Handbook of Medical Mycology.* 1980.)

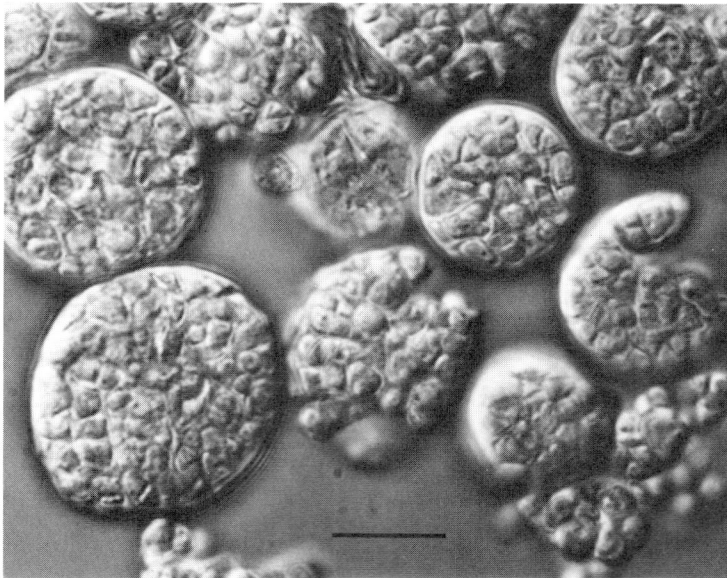

Figure 24. *Coccidioides immitis.* The spherules developed on modified Converse medium at 37°C. Bar is 10 μm.

Coccidioides
Rixford et Gilchrist, 1896

Description: Colonies are extremely variable and rapid growing, typically glabrous and grayish, later becoming white and cottony. With age, colonies become tan to brown in color at 25°C. Fertile hyphae usually arise at right angles to the vegetative hyphae, which are smaller in diameter. Arthroconidia are hyaline, 1-celled, rectangular to barrel shaped, alternating with empty disjunctor cells. Upon release by breakage of the disjunctor cells, the conidia have annular frills at both ends. At 37–40°C, large spherules containing endospores form on special media.

Salient characteristics: *Coccidioides immitis* can be recognized by the formation of alternating, hyaline arthroconidia in culture at 25–30°C, and by the formation of spherules containing endospores in tissue or at elevated temperatures on special media. *Coccidioides* differ from *Malbranchea* by forming spherules containing endospores under the appropriate conditions. A specific and sensitive exoantigen serologic procedure is available for the identification of cultures of *C. immitis.*

Laboratory precautions: *Coccidioides immitis* is an extremely dangerous fungus. It must be handled only in a biological safety cabinet.

Key references:
Huppert, Sun, and Rice, 1978
Sigler and Carmichael, 1976

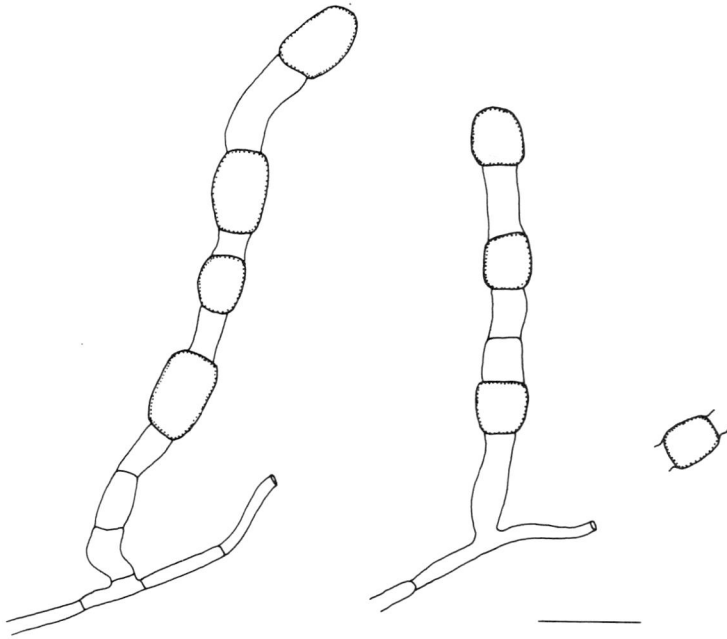

Coccidioides immitis

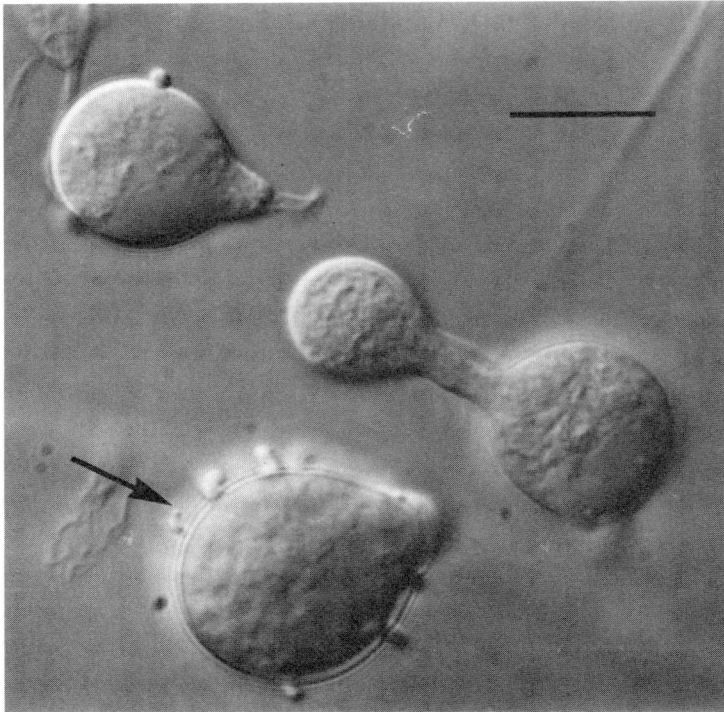

Figure 25. *Conidiobolus coronatus.* Spores and villose spores (arrow) can be seen in this preparation. Bar is 10 μm.

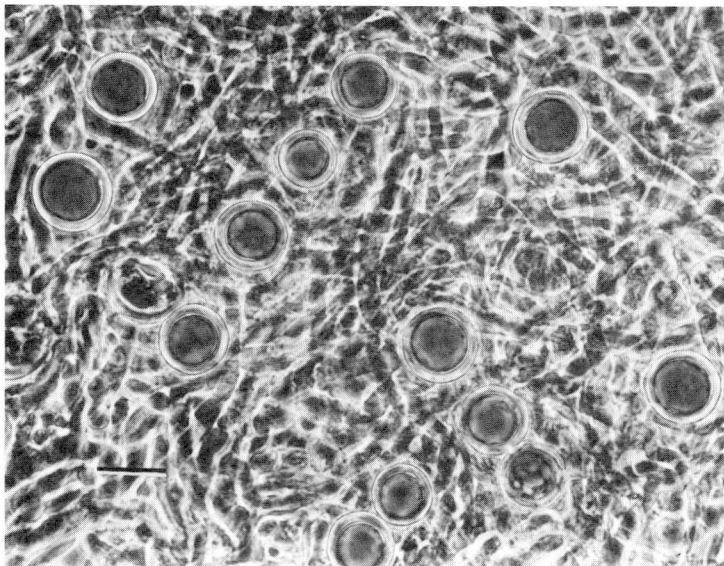

Figure 26. *Conidiobolus incongruus.* The zygospores do not have beaks associated with them. Bar is 10 μm.

Conidiobolus Brefeld, 1884

Description: Colonies are rapid growing, initially flat and glabrous, becoming folded irregularly and radially, white, with age becoming tan to brown in color. Primary spores are forcibly discharged, globose to pyriform with a single, broad, tapering basal papilla. Sporangiophores are hyphalike, not swollen at their apices. Globose replicative spores form from primary spores. Villose appendaged spores are produced by the direct conversion of the globose spores under dry conditions. Zygospores are usually globose to elongate, thick walled, formed in the larger of two gametangia, without beaks.

Salient characteristics: Species of *Conidiobolus* can be recognized by their rapid growth on routine media, forcibly discharged spores, replicative spores, and villose appendaged spores. *Conidiobolus* differs from *Entomophthora* by forming spherical to pyriform primary spores on unbranched sporangiophores, by having spores with four or more nuclei, by having a broadly rounded apex, by being cultured on routine media, and by having zygospores formed in the larger of two gametangia. *Conidiobolus* differs from *Basidiobolus* by having sporangiophores that are not swollen at their apices, by having primary spores with papillae, and by the absence of beaks associated with the zygospores.

Laboratory precautions: Handle with care, but special precautions are not necessary.

Key references:
Fuller, 1978
O'Donnell, 1979

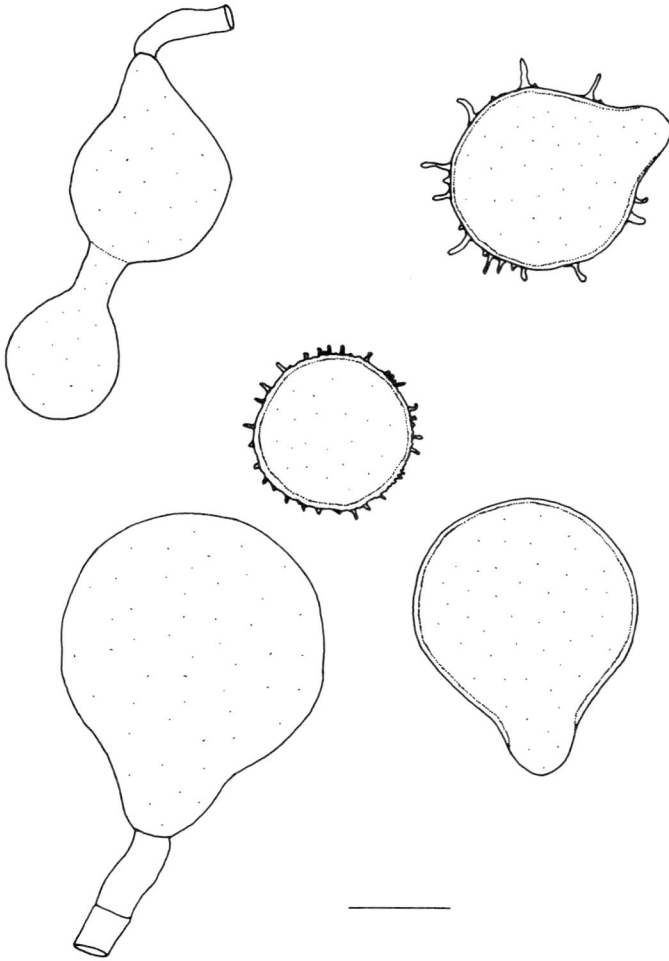

Conidiobolus coronatus

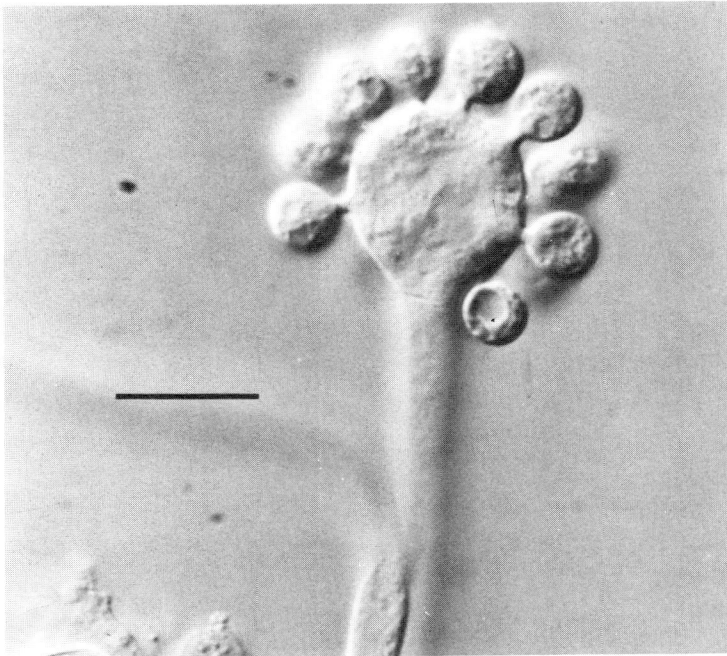

Figure 27. *Cunninghamella bertholletiae.* The spores develop from a vesicle at the apex of the sporangiophore. Bar is 10 μm.

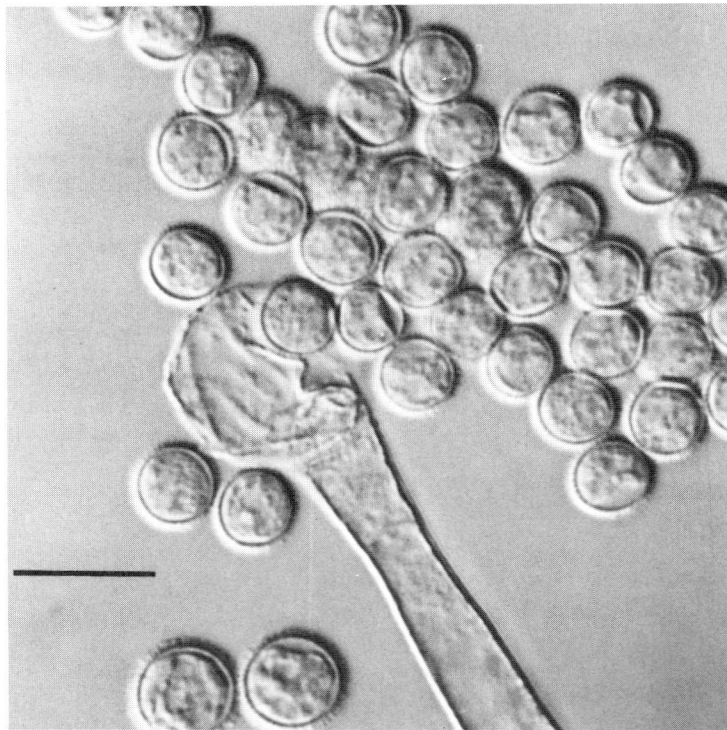

Figure 28. *Cunninghamella bertholletiae.* The spores are typically echinulate. Bar is 10 μm.

Cunninghamella
Matruchot, 1903

Description: Colonies are rapid growing, cottony, and white to dark gray in color. Sporangiophores are erect, branching, and sparsely septate, ending in a swollen vesicle. Spores are 1-celled, globose to ovoid, solitary, developing on swollen denticles from the vesicle surface. Spore walls often have needle-like crystals.

Salient characteristics: *Cunninghamella* species form 1-celled spores on denticles arising from the surface of terminal vesicles on branched sporangiophores. *Cunninghamella bertholletiae,* which has been frequently incorrectly identified as *C. elegans* in the clinical laboratory, is the only known human pathogen in the genus.

Laboratory precautions: Handle with care, but special precautions are not necessary.

Key reference:
Weitzman and Crist, 1979

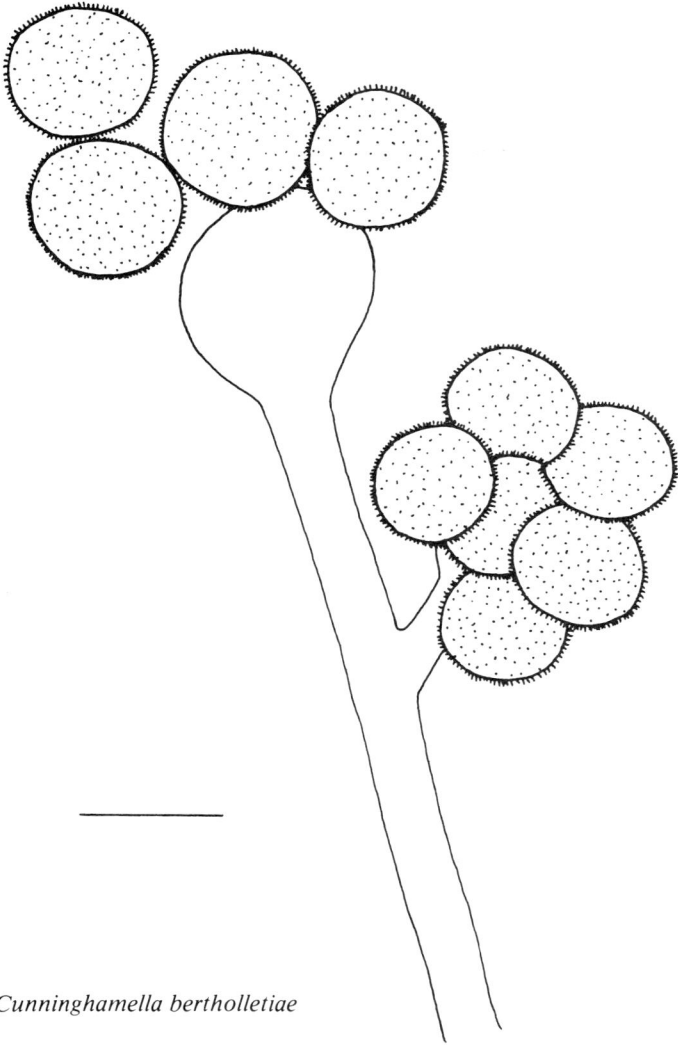

Cunninghamella bertholletiae

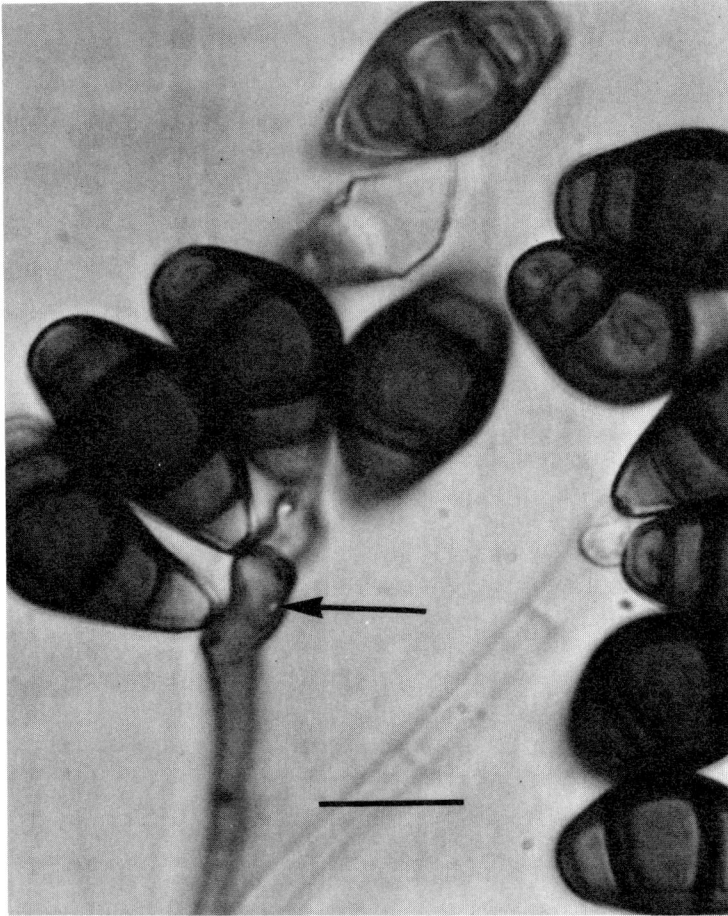

Figure 29. *Curvularia* sp. The poroconidia are curved with the end cells paler in color than the central cells. They arise from a geniculate conidiophore, in which a pore can be seen in the conidiophore wall (arrow). Bar is 10 μm.

Curvularia Boedijn, 1933

Description: Colonies are rapid growing, woolly, at first white, becoming dark olive green to black in color with a darkly pigmented reverse. Conidiophores are geniculate, septate, brown, branched or simple, solitary or in clusters, and sympodial. Conidia are several celled, pale brown to dark brown, with end cells lighter in color than central cells. The conidia typically are curved with an enlarged central cell, occasionally straight or pyriform, with a dark basal protuberant hilum.

Salient characteristics: Species of *Curvularia* produce geniculate, dark, sympodial conidiophores that bear curved, multicelled, brown conidia that have their end cells lighter in color than the central cell (the latter typically is the largest cell). *Curvularia* is readily distinguished from *Drechslera* by the fact that it has curved conidia with an enlarged central cell.

Laboratory precautions: Handle with care, but special precautions are not necessary.

Key references:
Ellis, 1971
Ellis, 1976
Somal, 1976

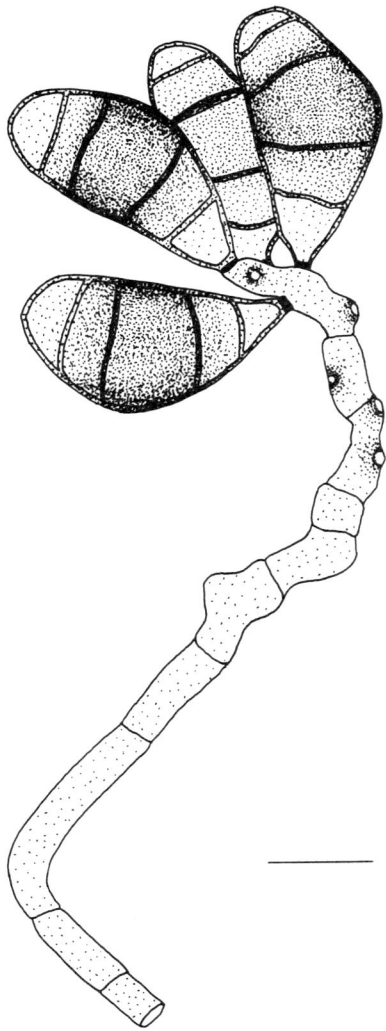

Curvularia sp.

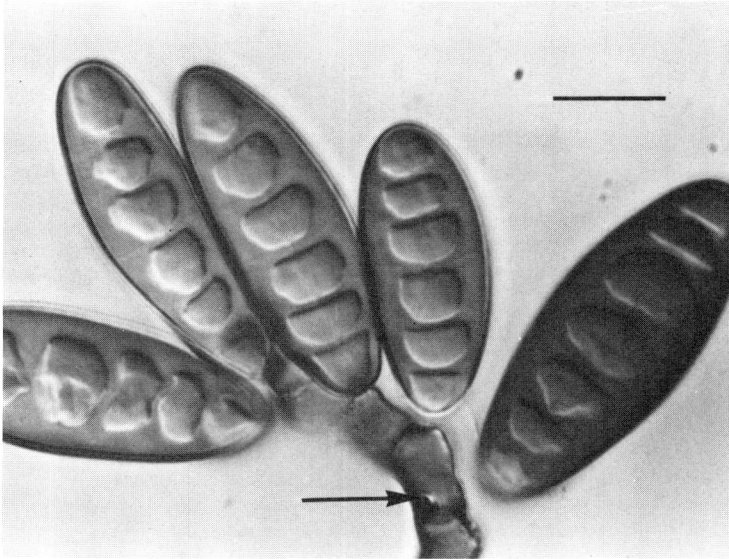

Figure 30. *Drechslera* sp. The poroconidia are fusoid and arise from a geniculate conidiophore. A pore can be seen in the conidiophore wall (arrow). Bar is 10 μm.

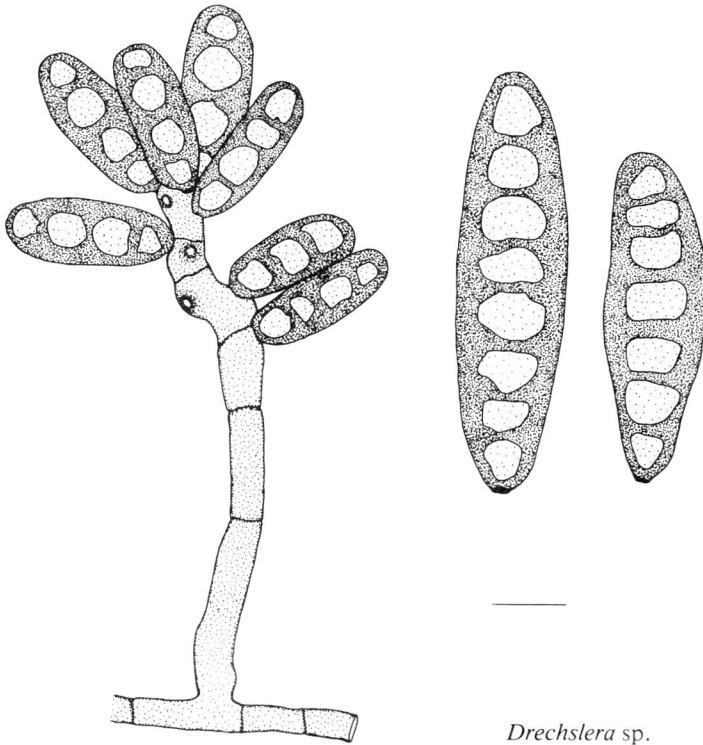

Drechslera sp.

Drechslera Ito, 1930

Description: Colonies are rapid growing, velvety to woolly, at first white, becoming olive green to black with a darkly pigmented reverse color. Conidiophores are sympodial, geniculate, brown, branched or simple. Conidia are multicelled, fusoid to cylindrical, light brown to dark brown, usually with a darkly pigmented hilum.

Salient characteristics: *Drechslera* species produce geniculate, pigmented, sympodial conidiophores that bear fusoid to cylindrical, multicelled, dematiaceous poroconidia. *Drechslera* differs from *Curvularia* by having fusoid to cylindrical conidia that are evenly pigmented; it differs from *Helminthosporium* by having geniculate, sympodial conidiophores.

Laboratory precautions: Handle with care, but special precautions are not necessary.

Key references:
Ellis, 1971
Ellis, 1976

58

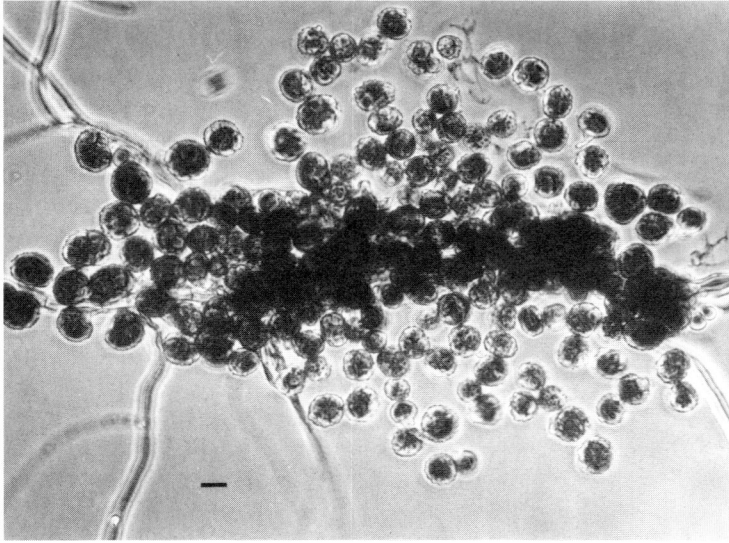

Figure 31. *Epicoccum nigrum.* The muriform dematiaceous conidia are forming on a sporodochium. Bar is 10 μm.

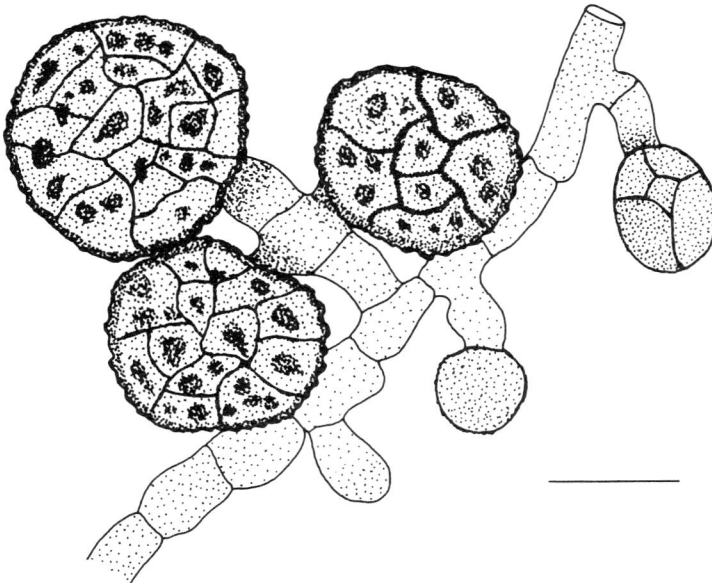

Epicoccum nigrum

Epicoccum Link ex Steudel, 1824

Description: Colonies are moderately rapid growing, cottony, later developing tufts containing brown to black sporodochia. Colony color is variable within each colony, white to pink, red, purple, yellow, and (rarely) olivaceous green or brown, with bright yellow-to-orange-to-purple colors dominant. A diffusible orange to brown pigment often colors the agar. Conidiophores are short. Conidia are dark brown, muriform, subglobose, large, with roughened or warty crusted cell walls.

Salient characteristics: Species of *Epicoccum* produce multicolored colonies that are often predominantly yellow to orange, with dark tufts containing sporodochia and with subglobose, brown, muriform conidia having roughened outer cell walls.

Laboratory precautions: Handle with care, but special precautions are not necessary.

Key reference:
Schol-Schwarz, 1959

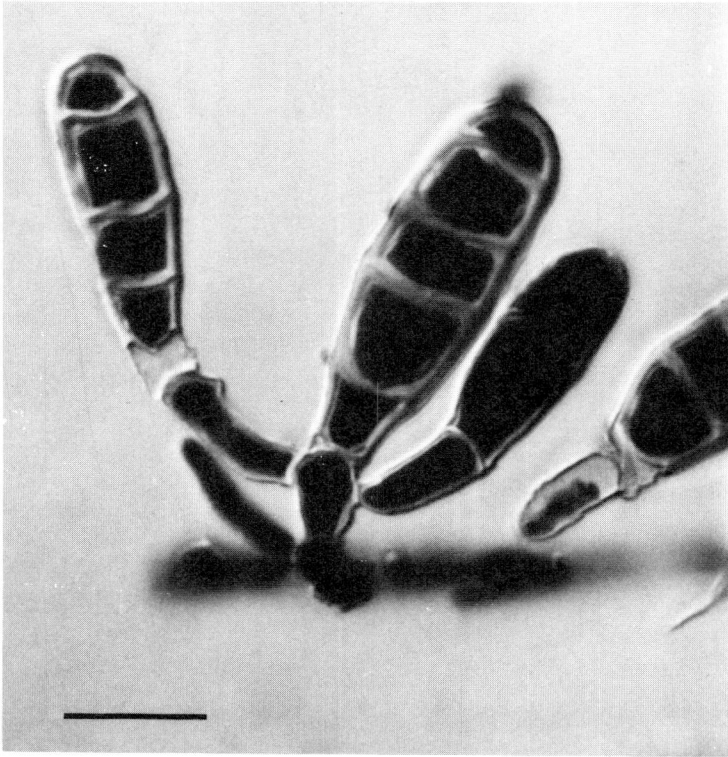

Figure 32. *Epidermophyton floccosum.* The conidia often occur in clusters. Bar is 10 μm.

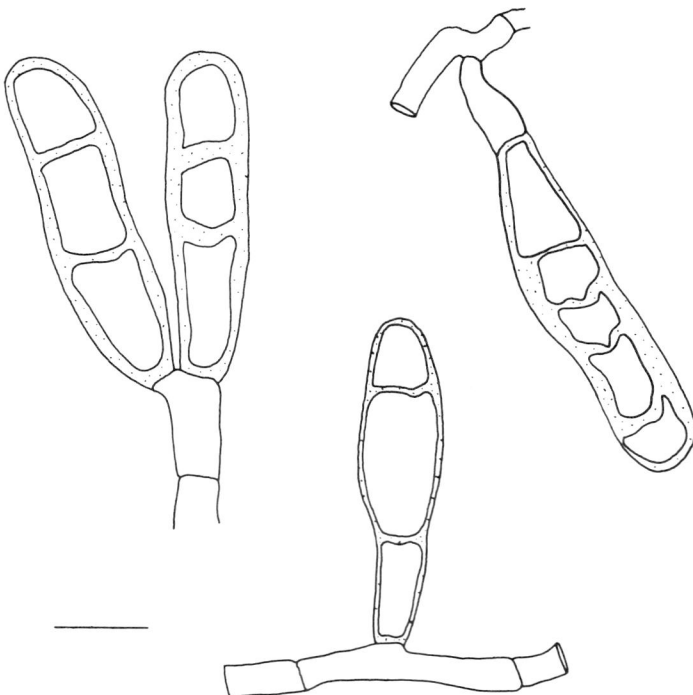

Epidermophyton floccosum

Epidermophyton
Sabouraud, 1907

Description: Colonies are slow growing, powdery and membranous, velvety, becoming woolly to suedelike, gently folded, khaki to olivaceous, pale yellow to yellow green in color with a yellowish tan to deep yellow brown reverse. The colonies quickly become downy and sterile. Conidia are thin walled, 3- to 5-celled, smooth, clavate, single or in clusters. Microconidia are absent. Chlamydoconidia are common in older cultures. *In vitro* hair perforation test is negative.

Salient characteristics: Isolates of *E. floccosum* are slow growing, velvety, khaki to olivaceous or yellow green, becoming suedelike and furrowed with age. They produce abundant, smooth-walled, 3- to 5-celled, clavate conidia that occur typically in clusters. *Epidermophyton floccosum* differs from *E. stockdaleae* by being unable to perforate hair.

Laboratory precautions: Handle with care, but special precautions are not necessary.

Key reference:
Rebell and Taplin, 1970

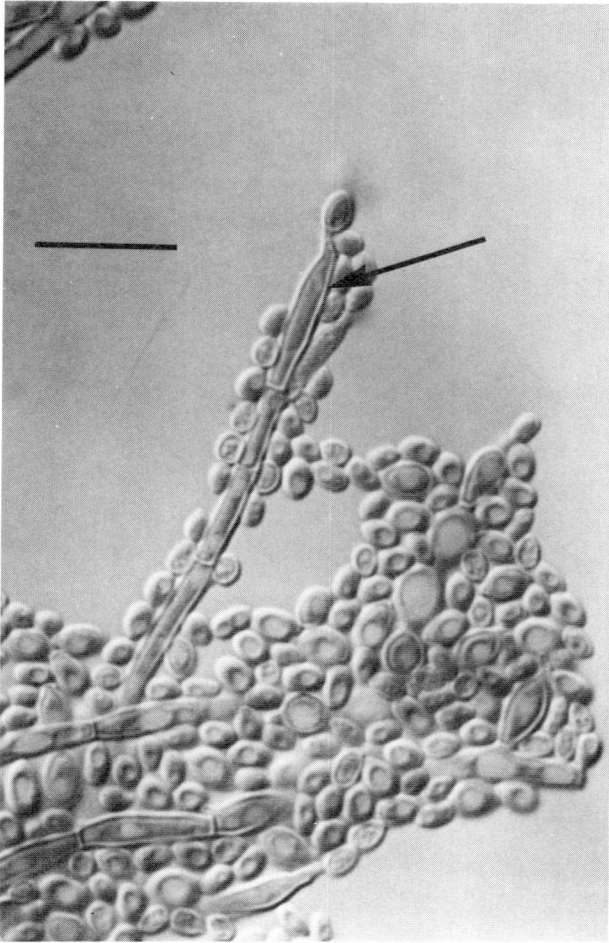

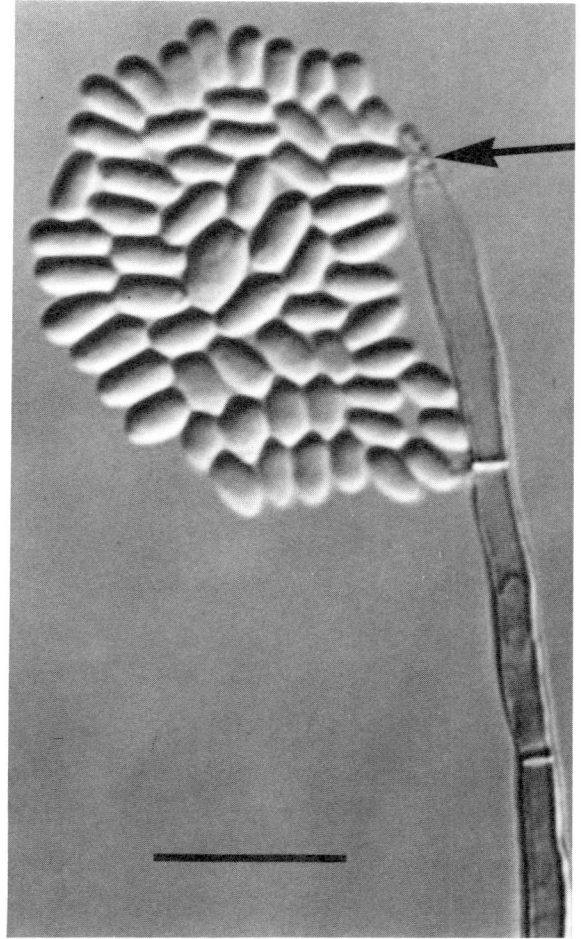

Figure 33. *Exophiala jeanselmei.* The annellides (arrow) of this species are cylindrical to lageniform. The yeasts belong to the genus *Phaeococcomyces.* Bar is 10 μm.

Figure 34. *Exophiala spinifera.* The annellations (arrow) can be seen at the apex of this annellide. Bar is 10 μm.

Exophiala Carmichael, 1966

Description: Colonies are slow to rapid growing, usually moist and yeastlike initially, becoming woolly with age, olivaceous gray to black in color. Annellides are cylindrical to flask shaped, tapering toward their apices, on hyphalike conidiophores, integrated within the hyphae, as yeast-like cells, or any combination. Conidia are 1-celled (several-celled in *E. salmonis* only), hyaline to pale brown, accumulating as a ball at the apex of the annellide.

Salient characteristics: Species of *Exophiala* are recognized by their cylindrical to lageniform annellides with balls of conidia at their apices. *Exophiala* species differ from members of the genera *Phialophora* and *Wangiella* by forming annellides rather than phialides.

61

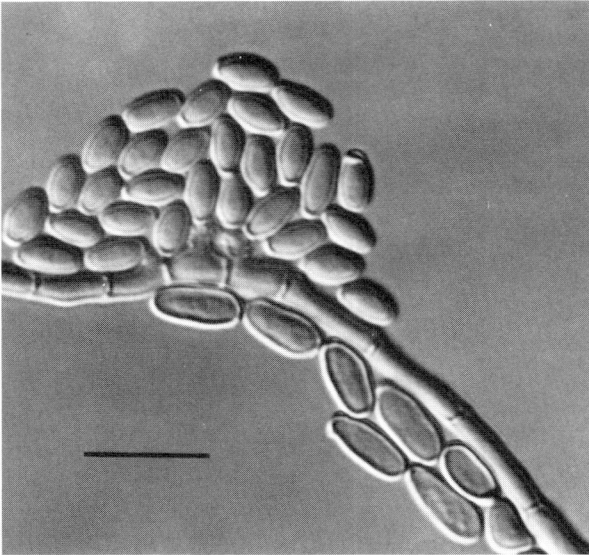

Figure 35. *Exophiala werneckii.* The 1-celled conidia are arising from intercalary annellides. Bar is 10 μm.

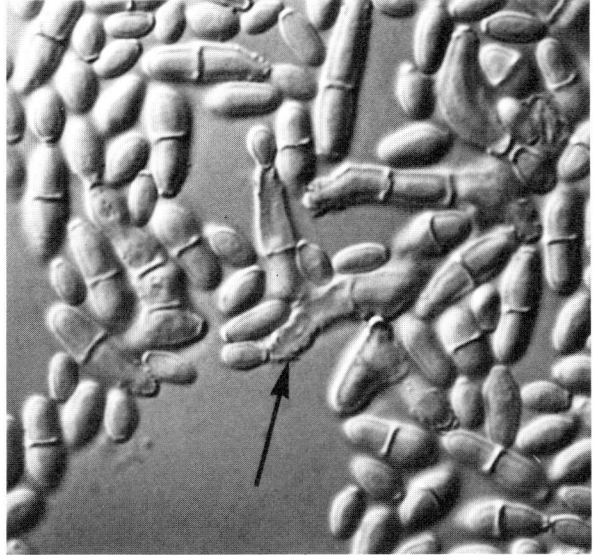

Figure 36. *Exophiala werneckii.* The yeast cells are typically 2-celled. Annellations (arrow) can be seen at the apices of some of the cells. (Reproduced by permission of Academic Press from M. McGinnis. *Laboratory Handbook of Medical Mycology,* 1980.)

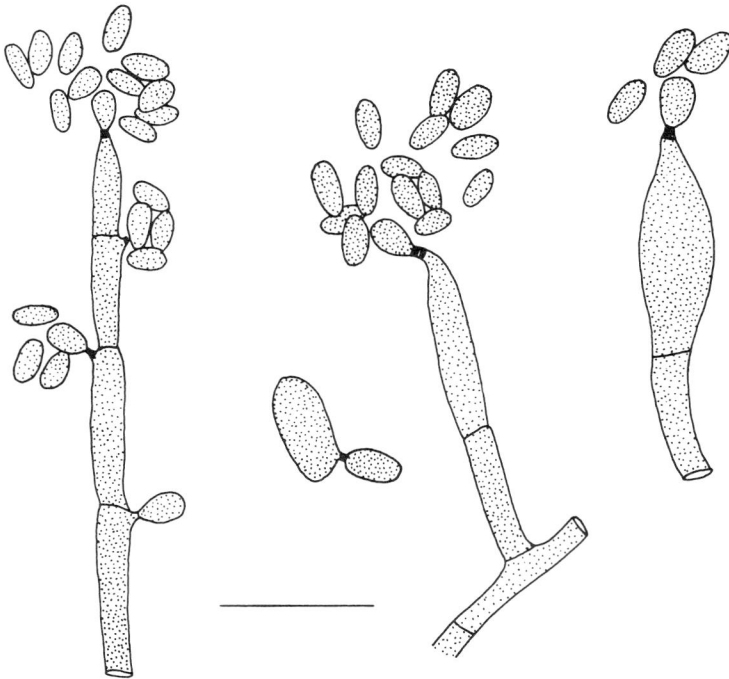

Exophiala jeanselmei

Key to the Human Pathogenic Species of *Exophiala*

1. Yeast cells 2-celled, usually the predominant form in the culture, rounded at one end, tapering toward the other with annellations at the tapering end; some annellides integrated within hyphae; annelloconidia 1-celled, occasionally 2-celled, accumulating in balls that tend to slide down the hyphae .... *E. werneckii*
1.´ 2-celled yeast cells absent 2
    2. Annellides arising on spinelike conidiophores from the hyphae .... *E. spinifera*
    2.´ Spinelike conidiophores absent 3
3. Annellides moniliform with long necks .... *E. moniliae*
3.´ Annellides cylindrical to lageniform .... *E. jeanselmei*

Laboratory precautions: Handle with care, but special precautions are not necessary.

Key reference:
McGinnis, 1980

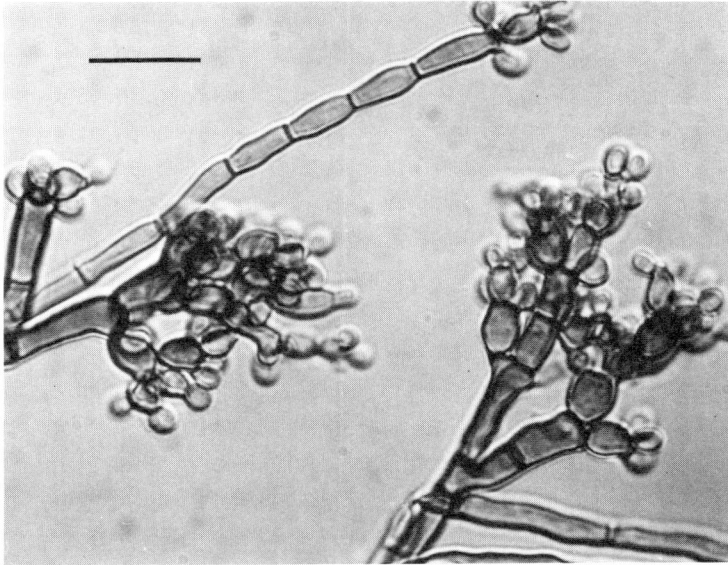

Figure 37. *Fonsecaea compacta.* The sympodial cells give rise to compact heads. Bar is 10 μm.

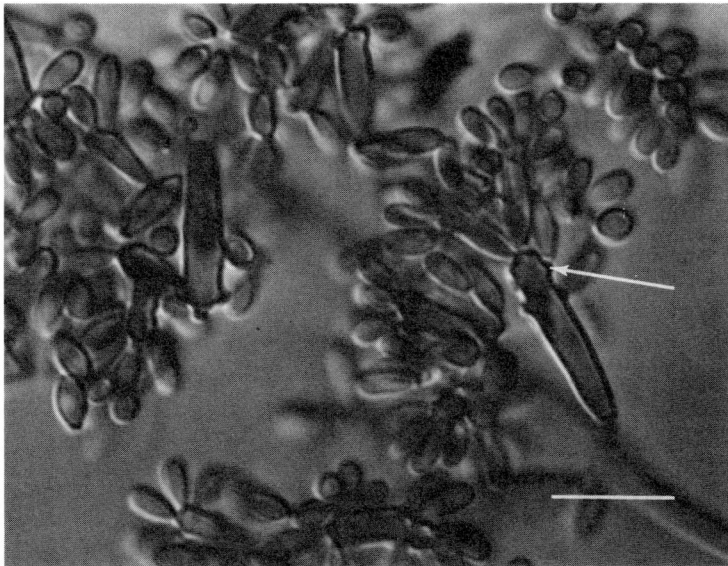

Figure 38. *Fonsecaea pedrosoi.* The apices of the sympodial cells are irregular with denticles bearing 1-celled conidia (arrow). Bar is 10 μm. (Reproduced by permission of Academic Press from M. McGinnis. *Laboratory Handbook of Medical Mycology,* 1980.)

Fonsecaea Negroni, 1936

Description: Colonies are slow to moderately rapid growing, flat to raised and folded, often brittle, velvety to cottony, olivaceous brown to olivaceous black in color. Conidiophores are pale brown, erect, septate, sympodial with conidiogenous zones confined to the upper portion. Conidia are 1-celled, arising upon swollen denticles. Primary conidia function as sympodial conidiogenous cells, becoming irregularly swollen at their apices, giving rise to 1-celled, pale brown, secondary conidia on swollen denticles. Secondary conidia often produce tertiary series of conidia like those formed by the primary conidia, resulting in a complex conidial head. Other types of associated conidia may include branching chains of dematiaceous conidia like those formed by the genus *Cladosporium,* dematiaceous phialides having collarettes like those produced by the genus *Phialophora,* and sympodial conidiophores bearing 1-celled, pale brown conidia like those seen in the genus *Rhinocladiella.*

Salient characteristics: *Fonsecaea* is recognized by its complex conidial head consisting of apically, irregularly swollen, series of conidia that function as conidiogenous cells. This type of development distinguishes *Fonsecaea* from *Rhinocladiella,* which produces a single row of 1-celled conidia around its conidiophore. *Fonsecaea compacta* differs from *F. pedrosoi* by having subglobose to globose conidia occurring in compact heads, in contrast to the loose heads consisting of elongated conidia in *F. pedrosoi.*

Laboratory precautions: Handle with care in a biological safety cabinet.

Key reference:
McGinnis, 1980

64

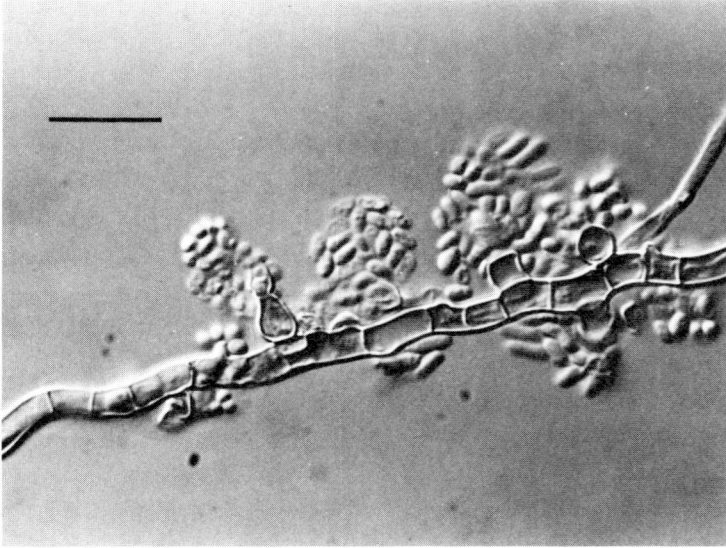

Figure 39. *Fonsecaea pedrosoi*. Isolates rarely produce phialides. Bar is 10 μm.

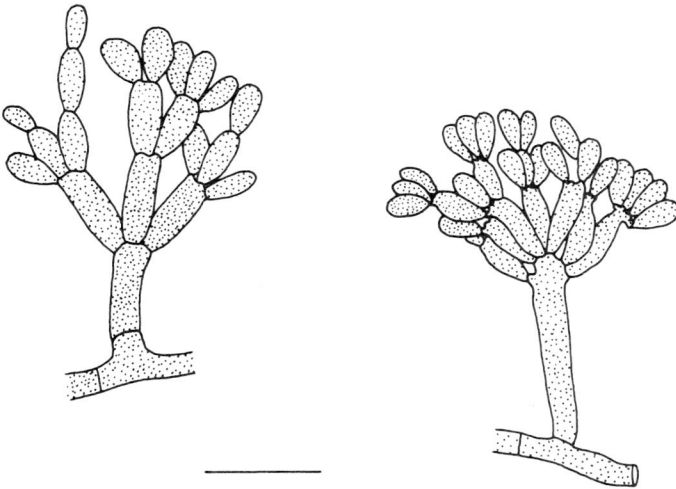

Fonsecaea pedrosoi

65

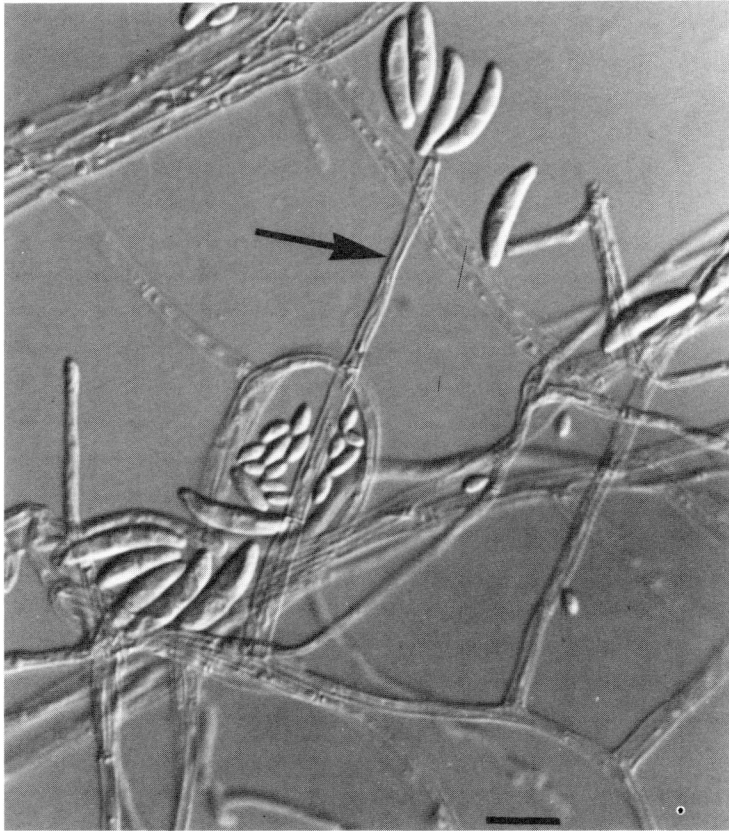

Figure 40. *Fusarium solani.* The curved macroconidia are accumulating at the apex of a phialide (arrow). Bar is 10 μm.

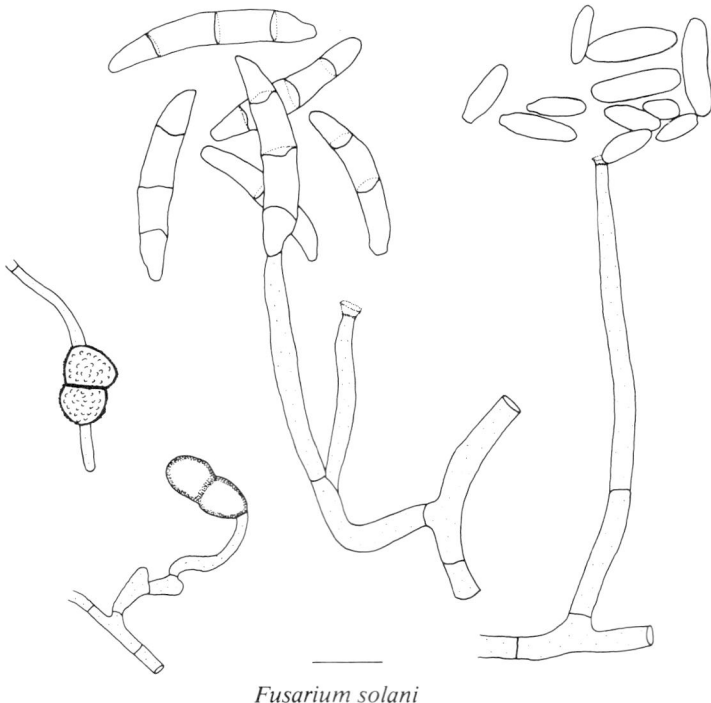

Fusarium solani

Fusarium Link ex Gray, 1821

Description: Colonies are rapid growing, woolly to cottony, flat, spreading, white, cream, tan, cinnamon, yellow, red to violet, or purple. Phialides are cylindrical, with a small collarette, solitary or as a component of a complex branching system. Macroconidia are 2- or more celled, smooth, cylindrical to curved, with a distinct basal foot cell, and tend to accumulate in balls or rafts. Microconidia are 1-celled (occasionally 2-celled), smooth, hyaline, ovoid to cylindrical, accumulating in balls (occasionally occurring in chains). Sporodochia are usually absent in culture.

Salient characteristics: Members of the genus *Fusarium* produce sickle-shaped, multicelled macroconidia with foot cells. *Fusarium* differs from *Cylindrocarpon* by having macroconidia with foot cells and pointed distal ends; it differs from *Acremonium* by having macroconidia.

Laboratory precautions: Handle with care, but special precautions are not necessary.

Key references:
Booth, 1971
Booth, 1977

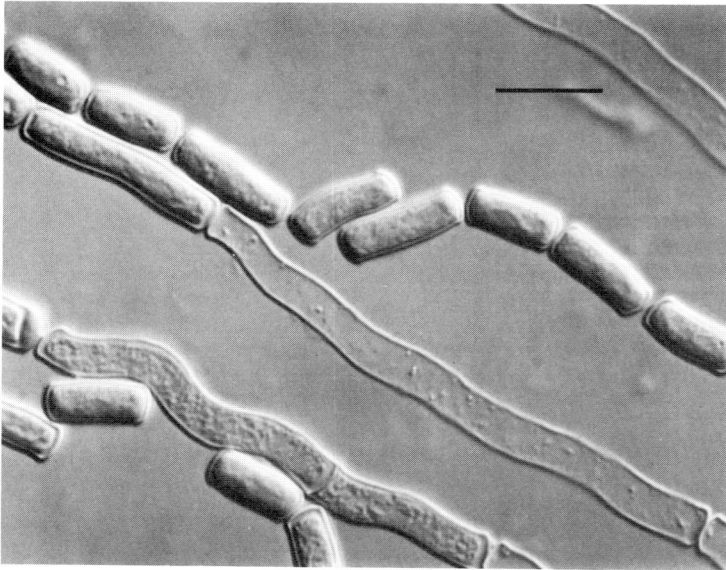

Figure 41. *Geotrichum candidum.* The hyphae are fragmenting into 1-celled arthroconidia. Bar is 10 μm.

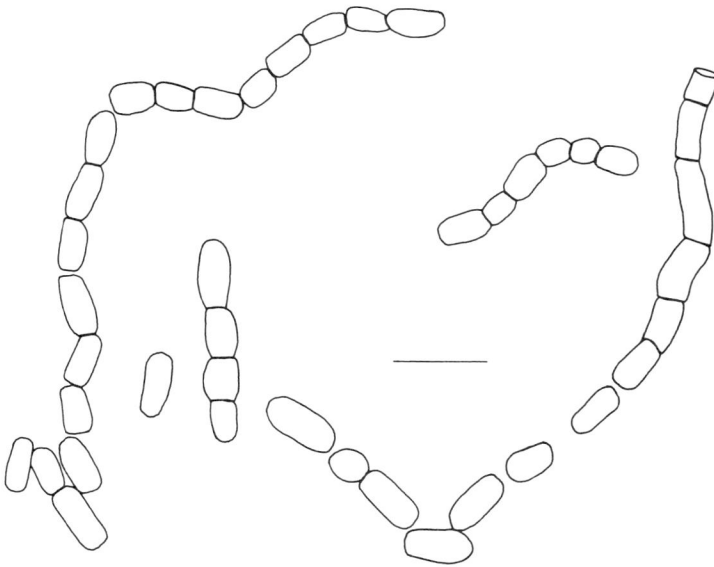

Geotrichum candidum

Geotrichum Link ex Persoon, 1822

Description: Colonies are rapid growing, dry, powdery to cottony (becoming yeastlike or slimy when the colony surface is disturbed), white in color. Conidiophores are absent. Arthroconidia are 1-celled, in chains, hyaline, and result from the fragmentation of undifferentiated hyphae by fission through double septa. See Table 2.3 for physiologic characteristics.

Salient characteristics: The genus *Geotrichum* is characterized by chains of hyaline arthroconidia developing from undifferentiated hyphae that are released by fission through double septa. *Geotrichum* differs from *Scytalidium* by having hyaline arthroconidia; from *Arthrographis* and *Oidiodendron* by lacking conidiophores; from *Malbranchea* by having fission arthroconidia rather than arthroconidia released by disjunctor cells; and from *Trichosporon* and *Moniliella* by lacking blastoconidia. *Geotrichum* contains two species of medical interest: *G. candidum* and *G. penicillatum*.

Laboratory precautions: Handle with care, but special precautions are not necessary.

Key references:
Gueho and Buissière, 1975
Weijman, 1979

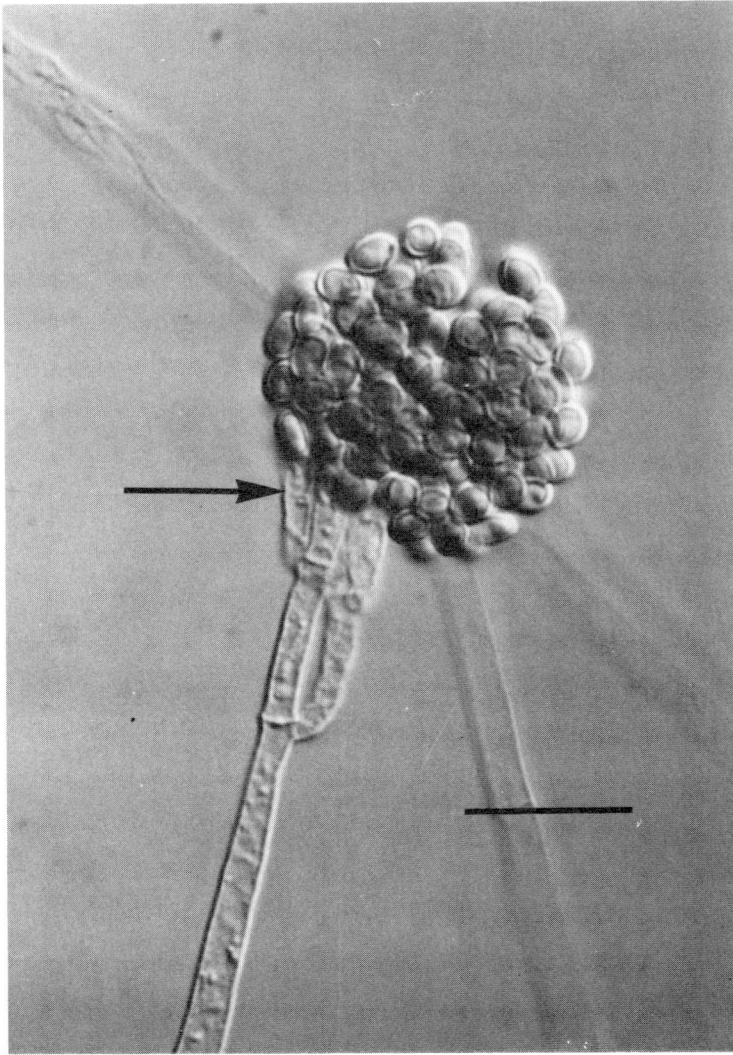

Figure 42. *Gliocladium atrum.* The conidia accumulate as a single ball at the apices of the phialides (arrow). Bar is 10 μm.

Gliocladium Corda, 1840

Description: Colonies are rapid growing, spreading, cottony, white to cream, pink to rose, or dark green, with a colorless, white, or yellowish reverse. Conidiophores are erect and branch repeatedly at their apices. The terminal branches give rise to flask-shaped phialides. Conidia are 1-celled, ovoid to cylindrical, accumulating in a single, terminal, large ball, or occasionally in a loose column.

Salient characteristics: A penicillus bearing a single, large, slimy ball of 1-celled conidia is typical of the genus *Gliocladium*.

Laboratory precautions: Handle with care, but special precautions are not necessary.

Key reference:
Barron, 1968

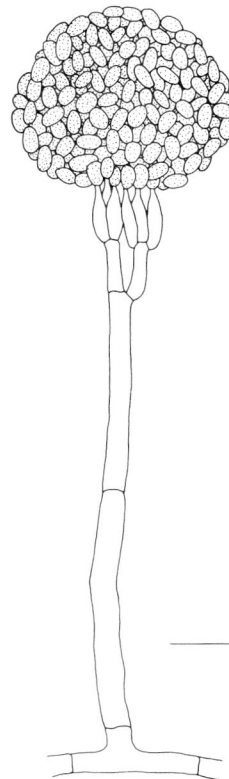

Gliocladium sp.

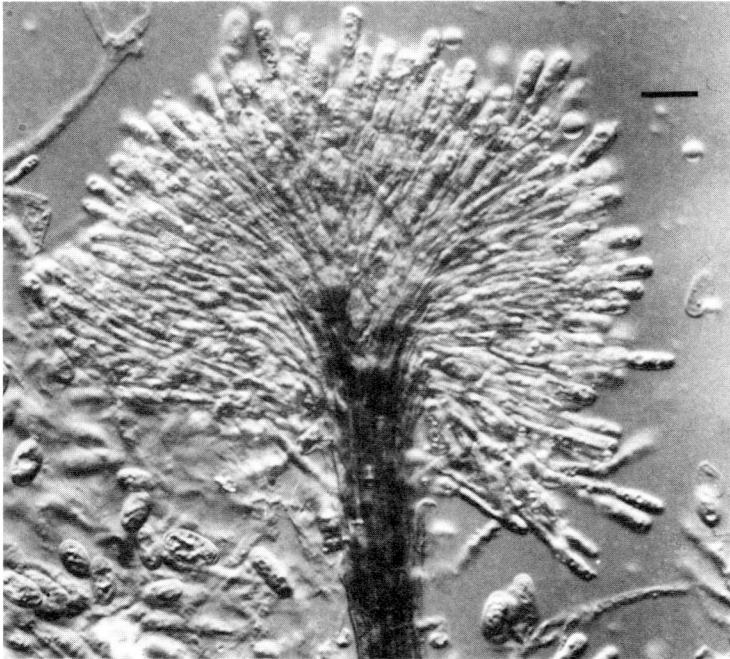

Figure 43. *Graphium* sp. The annelloconidia accumulate as a single ball at the apex of the synnema. Bar is 10 μm.

Graphium Corda, 1837

Description: Colonies are moderately rapid growing, woolly to cottony, gray, with synnemata that are erect, solitary or in clusters, and darkly pigmented. The conidiogenous cells are annellides that occur at the apices of the synnemata. Conidia are 1-celled, hyaline, smooth, subglobose to ovoid, accumulating as a single large ball at the apex of the synnema.

Salient characteristics: *Graphium* species are recognized by their distinctive, erect, black synnemata, each bearing a single, terminal, ball of 1-celled, hyaline conidia produced from annellides. *Graphium* must be distinguished from *Pesotum*, which produces sympodial conidiophores; and also from *Phialographium,* which produces phialides.

Laboratory precautions: Handle with care, but special precautions are not necessary.

Key reference:
Upadhyay and Kendrick, 1974

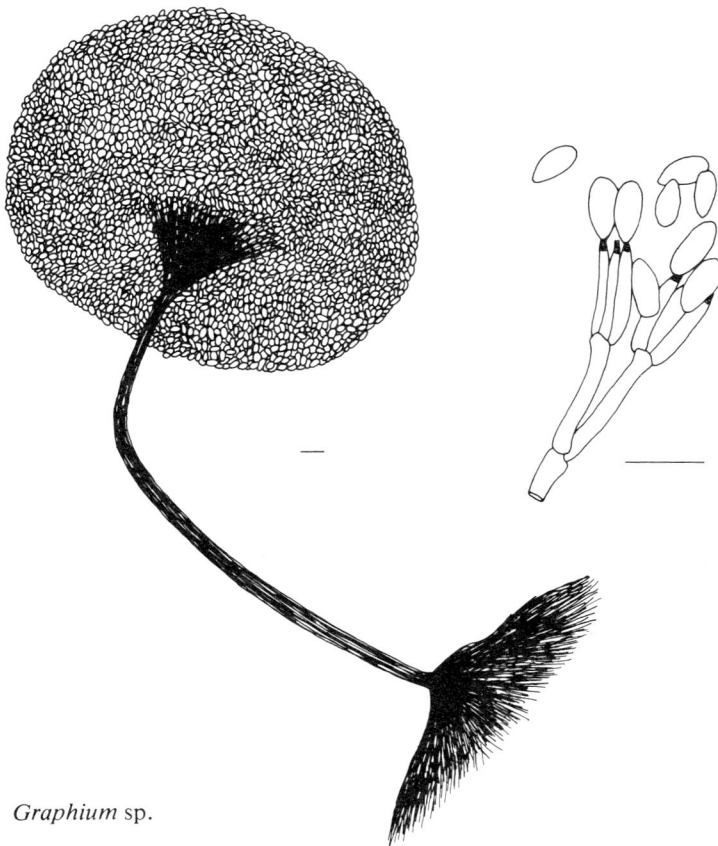

Graphium sp.

69

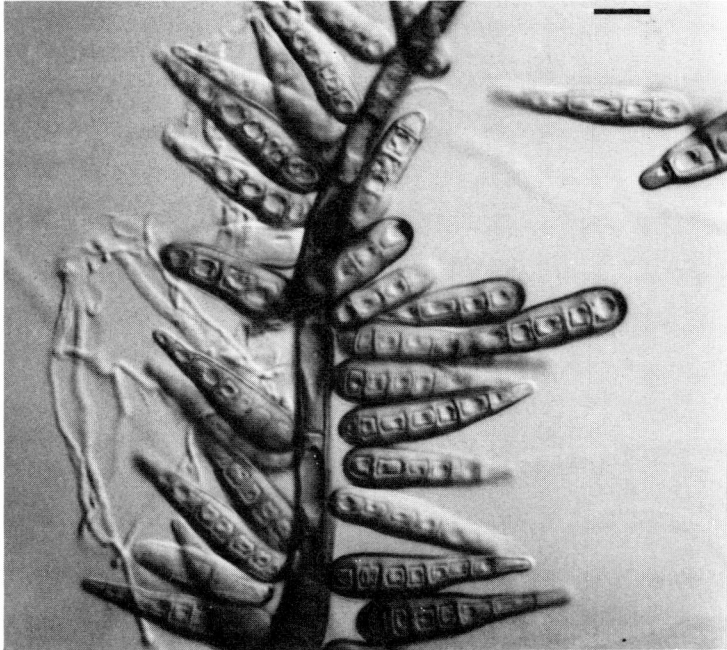

Figure 44. *Helminthosporium solani.* The obclavate conidia formed along the parallel-walled conidiophore. Bar is 10 μm.

Helminthosporium Link ex Fries, 1821 *nom. cons.*

Description: Colonies are rapid growing, velvety to woolly, olive green to black in color. Conidiophores are brown to dark brown, erect, ceasing to elongate when the terminal conidium is formed. Conidia are several celled, solitary, obclavate, pale to dark brown, occurring around the conidiophore.

Salient characteristics: *Helminthosporium* produces obclavate poroconidia from parallel-walled dematiaceous conidiophores that cease to increase in length after the terminal conidium is formed. *Helminthosporium* differs from *Drechslera* by forming parallel-walled, determinate conidiophores.

Laboratory precautions: Handle with care, but special precautions are not necessary.

Key reference:
McGinnis, 1980

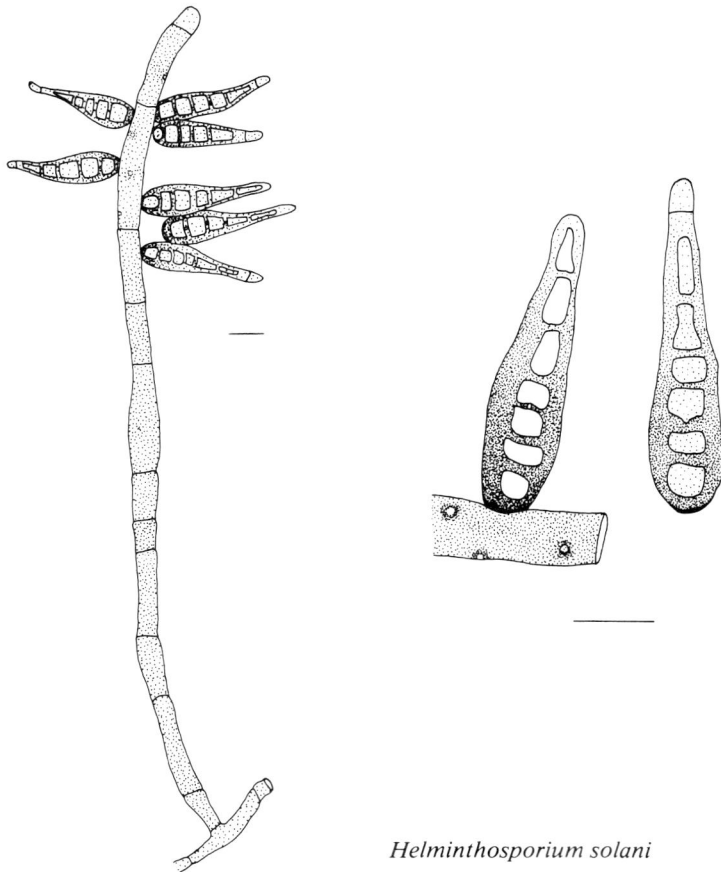

Helminthosporium solani

70

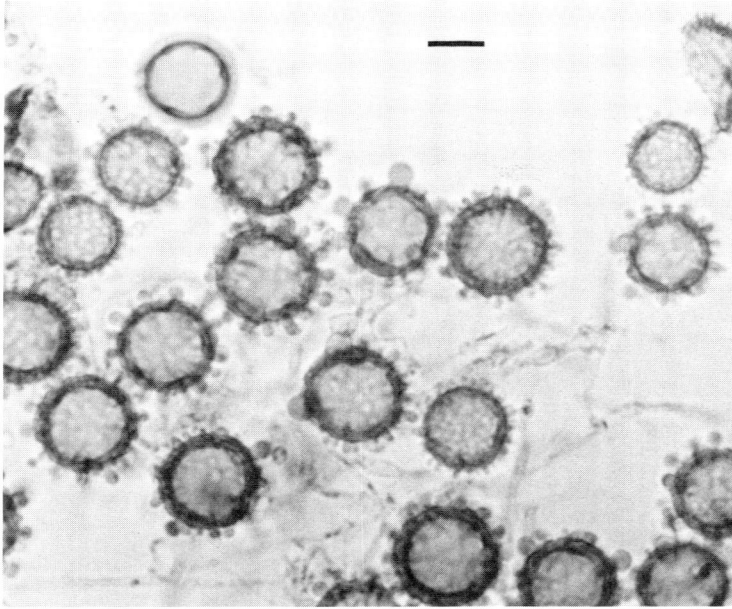

Figure 45. *Histoplasma capsulatum.* The macroconidia have coglike projections from their cell walls, 25°C. Bar is 10 μm. (Reproduced by permission of Academic Press from M. McGinnis. *Laboratory Handbook of Medical Mycology*, 1980.)

Histoplasma Darling, 1906

Description: Colonies are slow growing, granular to cottony, white initially, usually becoming buff brown with age, often with a yellow or yellowish orange reverse color. Conidiophores are hyphalike, arising at right angles to the parent hyphae. Macroconidia are 1-celled, large, thick walled, smooth or tuberculate, often with fingerlike projections. Microconidia are 1-celled, hyaline, smooth or echinulate, small. The fungus is dimorphic, growing as a small, ovoid, budding yeast at 37°C upon brain heart infusion agar containing 5% blood.

Salient characteristics: Isolates of *H. capsulatum* at 25°C grow slowly and have white to brown colonies with thick-walled, tuberculate, 1-celled macroconidia arising from hyphalike conidiophores; at 37°C colonies form small, 2–5 μm, narrow-based, ovoid, budding yeast cells. *Histoplasma* differs from *Chrysosporium* and *Sepedonium* by producing a yeast form at 37°C and by having specific exoantigens.

Laboratory precautions: Isolates of *Histoplasma* must be handled with caution in a biological safety cabinet.

Key reference:
McGinnis, 1980

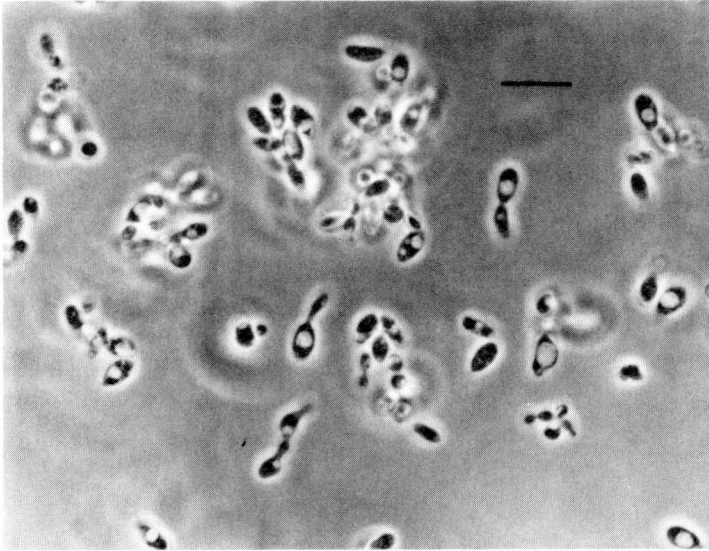

Figure 46. *Histoplasma capsulatum*. The yeast form was produced at 37°C on an enriched medium. Bar is 10 μm.

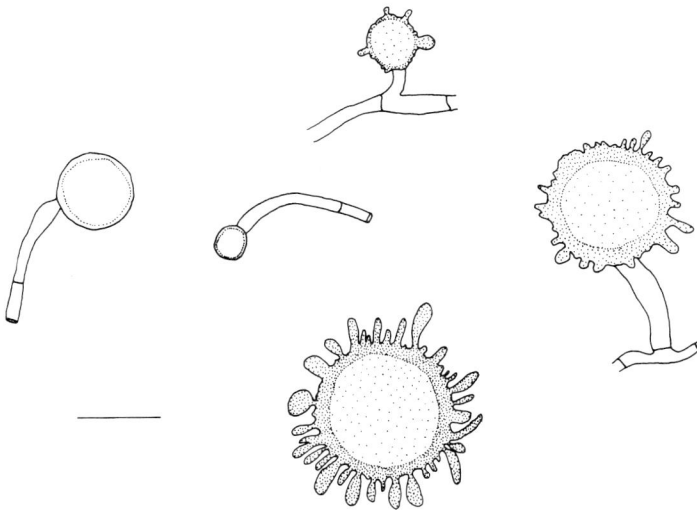

Histoplasma capsulatum

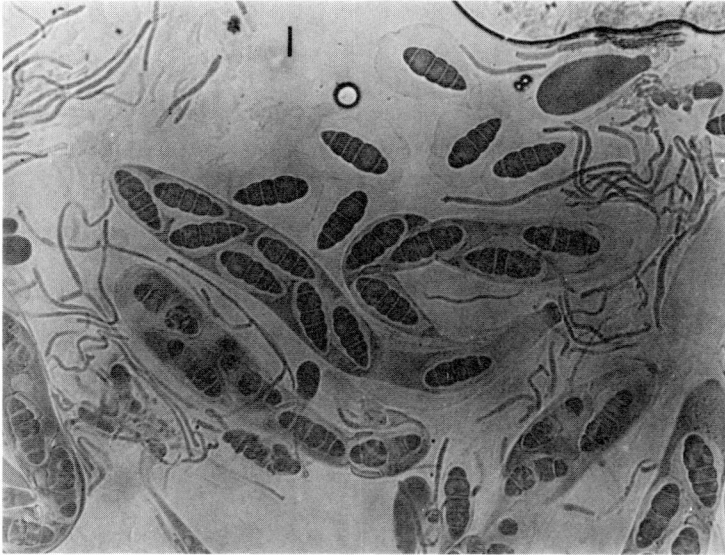

Figure 47. *Leptosphaeria senegalensis.* The ascospores are predominantly 5-celled. Bar is 10 μm. (Reproduced by permission of A. El-Ani and *Mycologia* from *Mycologia* 57:275–78, 1965.)

Leptosphaeria Cesati et de Notaris, 1861

Description: Colonies are slow growing, woolly, dark olive with a gray margin and with a dark olive to black reverse that is surrounded by a grayish margin. Ascostromata are without ostioles, globose to subglobose, and black. The asci are clavate to cylindrical, 8-spored, and bitunicate. Ascospores are 4- to 9-celled, hyaline or pigmented, fusoid to curved, with a constriction at each septum.

Salient characteristics: *Leptosphaeria* species form black, globose to subglobose ascostromata containing 8-spored asci with 4- to 9-celled, hyaline to dark, fusoid to curved ascospores having constrictions at their septa. *Leptosphaeria tompkinsii* forms larger ascocarps, asci, and ascospores than does *L. senegalensis*.

Laboratory precautions: Handle with care, but special precautions are not necessary.

Key references:
El-Ani and Gordon, 1965
El-Ani, 1966

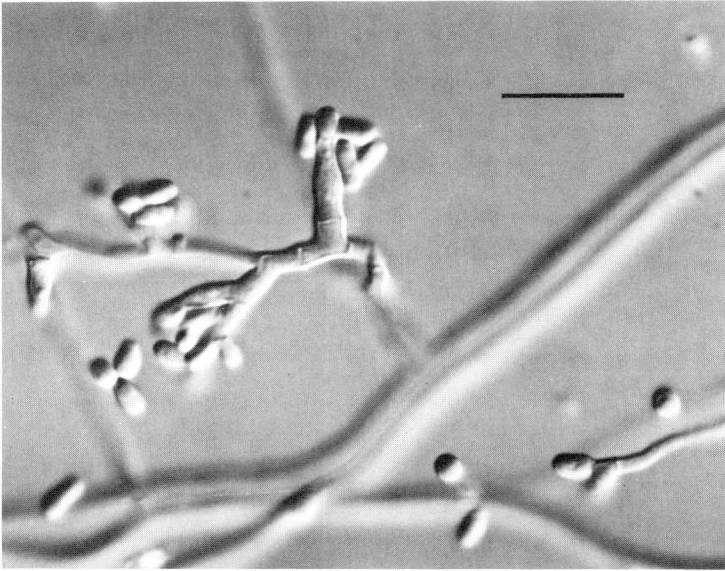

Figure 48. *Madurella mycetomatis.* Some isolates produce phialides. Bar is 10 μm.

Madurella Brumpt, 1905

Description: Colonies are variable, slow growing, raised to heaped, sometimes radially folded, woolly, and olivaceous to dark gray, yellow, or brown in color, sometimes producing a brownish diffusible pigment. Colonies are sterile, but sclerotia may be formed.

Salient characteristics: Isolates of *Madurella* are recovered from cases of black-grained eumycetoma. In the laboratory they are dematiaceous and sterile. Under some conditions, occasional isolates of *M. mycetomatis* form phialides having collarettes. *Madurella mycetomatis* grows well at 37°C; *M. grisea* does not.

Laboratory precautions: Handle with care, but special precautions are not necessary.

Key reference:
McGinnis, 1980

Malbranchea Saccardo, 1882

Description: Colonies are moderately rapid growing, raised or flat, with or without furrows, powdery, woolly, or cottony, and white, orange, buff, tan, brown, or dark golden in color. Zones of different colors may be present. Conidiophores are absent. Arthroconidia are 1-celled, cylindrical, straight or curved, truncate, hyaline to greenish-yellow, not wider in diameter than the hypha that bears them, alternate with disjunctor cells, and are released by the lysis or fracture of the disjunctor cells.

Salient characteristics: Nondematiaceous, 1-celled, alternating arthroconidia that are no wider than the hyphae that bear them are characteristic of *Malbranchea* species. *Malbranchea* differs from *Coccidioides immitis* by failure to produce spherules containing endospores, and by not reacting with *C. immitis*-specific reagents in the exoantigen test.

Laboratory precautions: Due to the extreme similarity of *Malbranchea* and *C. immitis,* isolates should be studied only in a biological safety cabinet until they are identified.

Key reference:
Sigler and Carmichael, 1976

Microsporum Gruby, 1843

Description: Colonies are slow to rapid growing, glabrous to cottony, white to brightly colored. Conidiophores are hyphalike. Macroconidia are 2- to several celled, thin to thick walled, echinulate to roughened, solitary, obovoid in one species, typically spindle shaped, hyaline, often with an annular frill. Microconidia are 1-celled, smooth, thin walled, hyaline, ovoid to clavate, solitary.

Salient characteristics: Isolates of *Microsporum* produce macroconidia that are spindle shaped to elliptic, thin to thick walled, echinulate to roughened. These isolates also produce 1-celled, smooth, hyaline, ovoid to clavate microconidia. *Microsporum* differs from *Trichophyton* and *Epidermophyton* by having echinulate to roughened macroconidia.

Laboratory precautions: Handle with care, but special precautions are not necessary.

Key references:
Dvořák and Otcenásek, 1969
Rebell and Taplin, 1970

Key to the More Common Species of *Microsporum*

1. *In vitro* hair perforation test negative 2
1.´ *In vitro* hair perforation test positive 4
    2. Diffusible strawberry-red pigment produced; macroconidia clavate to cigar shaped, with walls that are smooth to finely echinulate .... *M. gallinae*
    2.´ Diffusible strawberry-red pigment absent; macroconidia absent (or if present, irregular and contorted) ... 3
3. Colonies slow growing with many folds, glabrous to membranous, sometimes velvety, yellow to deep rust colored with a yellow to dull orange reverse; some hyphae often thick walled and highly segmented; usually nonsporulating, but forming fusiform macroconidia on rice grains or dilute Sabouraud dextrose agar .... *M. ferrugineum*
3.´ Colonies slow growing, glabrous to velvety, whitish tan to brownish, reverse usually salmon in color; terminal vesicles with solitary spinelike projections are often present .... *M. audouinii*
    4. Macroconidia irregular, contorted .... *M. distortum*
    4.´ Macroconidia not contorted 5
5. Macroconidia typically 2- to 3-celled, obovoid, thin walled and rough .... *M. nanum*
5.´ Macroconidia more than 3-celled, spindlelike or elliptic in shape 6
    6. Colonies deep yellow or buff to cinnamon brown 7
    6.´ Colonies white, yellowish tan, pale rose, or deep rose ... 8
7. Macroconidia spindle shaped, often with a predominant apical knob, with outer cell walls thicker than septal walls; colonies typically have a white to deep yellow color with a yellow to orange reverse .... *M. canis*
7.´ Macroconidia elliptic with thin outer cell and septal walls, outer cell walls tend to collapse slightly between the septa; distinct annular frill present; colonies granular to powdery, buff to cinnamon brown .... *M. gypseum*
    8. Colonies yellowish, greenish buff to dark brown with red reverse color; macroconidia thick walled, 6- to 10-celled, 12–28 × 31–75 µm .... *M. cookei*
    8.´ Colonies white to yellowish, tan or pink to deep rose, light yellow reverse color; macroconidia 5- to 12-celled, 8–13 × 44–88 µm, cylindrical to cigar shaped ... *M. vanbreuseghemii*

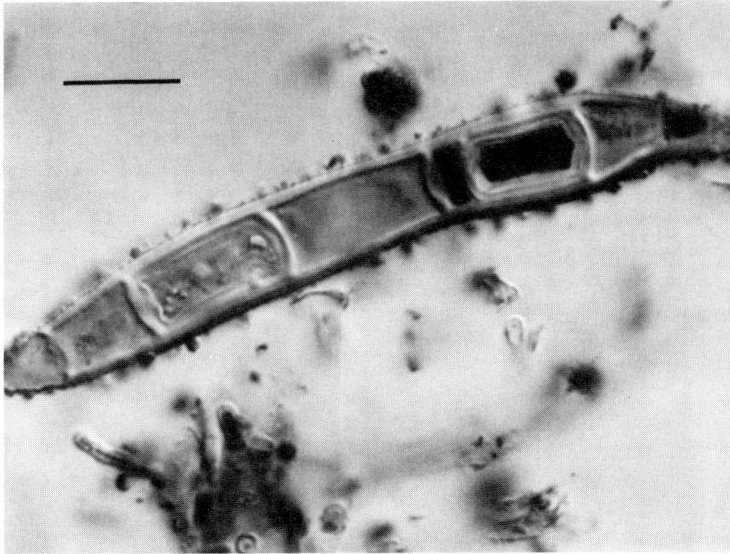

Figure 49. *Microsporum audouinii.* This rare macroconidium is irregular in shape. Bar is 10 μm.

Microsporum audouinii Gruby, 1843

Description: Colonies are slow growing, flat, spreading, dense, with a furlike mat having radiating edges, grayish white to tannish white, rarely rust, with a reverse color of salmon pink to peach or rose brown. Macroconidia are rare; when present, they are smooth to sparsely echinulate, poorly developed, thick walled, irregularly spindle shaped. Microconidia are rare; when present, they are 1 celled and ovoid to clavate. *In vitro* hair perforation test is negative. Terminal vesicles often have short spinelike solitary projections.

Salient characteristics: Characteristics of *M. audouinii* include flat, slow growing, downy colonies that are colorless, gray or tan, having terminal vesicles and pectinate hyphae. *Microsporum audouinii* differs from *M. canis* by neither perforating hair nor growing on polished rice grains.

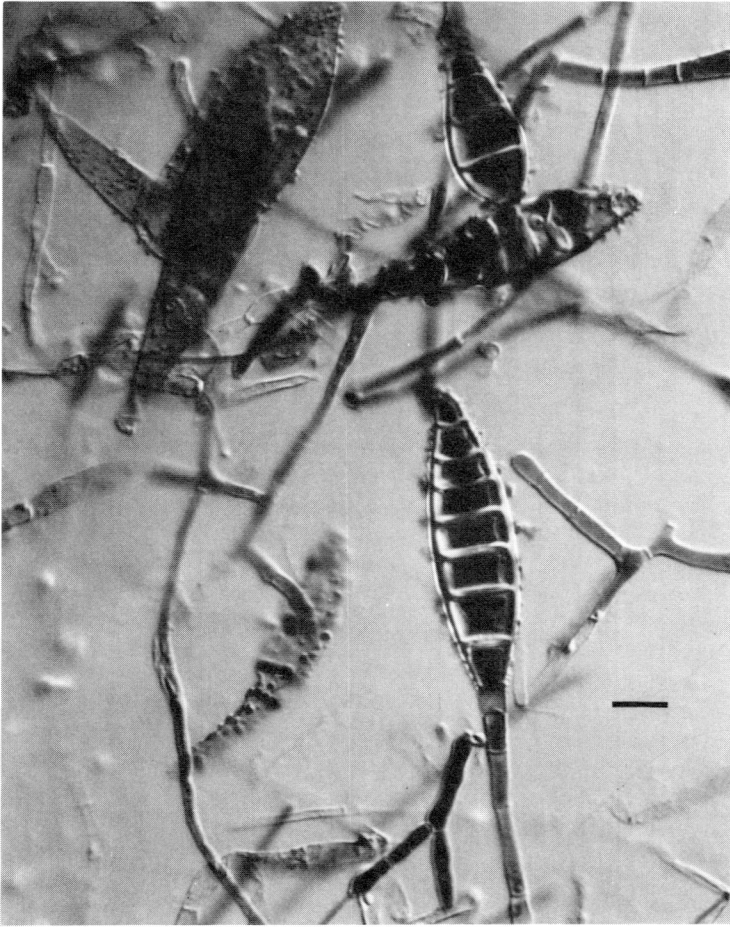

Microsporum canis Bodin, 1902

Description: Colonies are rapid growing, woolly to cottony, flat to sparsely grooved, white to yellowish, with a deep chrome yellow reverse color. Macroconidia are 6- to 15-celled, long, rough, with thick outer cell walls and thin septal walls, spindle shaped, and with an asymmetric apical knob. Microconidia are 1-celled and clavate to pyriform. *In vitro* hair perforation test is positive.

Salient characteristics: *Microsporum canis* is characterized by producing white to yellow, woolly to granular, rapid growing, flat colonies having roughened, spindle-shaped macroconidia with thickened outer cell walls and thin septal walls, and an annular frill. *Microsporum canis* differs from *M. audouinii* by perforating hair and growing on polished rice grains. On the rice grains, a deep yellow pigment typically is produced.

Figure 50. *Microsporum canis.* The septal wall of the macroconidia are thinner than the thick outer cell walls. Bar is 10 μm.

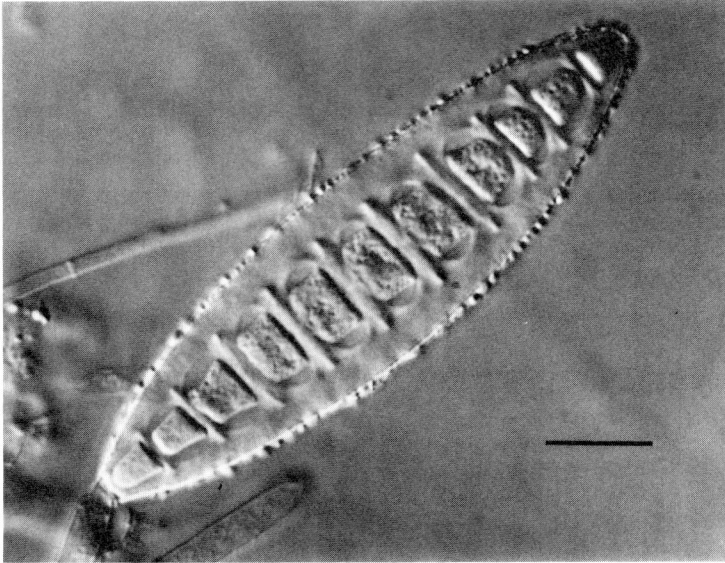

Microsporum cookei Ajello, 1959

Description: Colonies are moderately fast growing, powdery to coarse, spreading, becoming grape red, yellowish, greenish buff, or dark brown. Macroconidia 6- to 10-celled, ellipsoid, thick walled, and rough. Microconidia are 1-celled and ovoid to pyriform. *In vitro* hair perforation test is positive.

Salient characteristics: *Microsporum cookei* has coarse, grape red, yellow to dark brown spreading colonies with thick-walled, roughened, 6- to 10-celled, ellipsoid macroconidia.

Figure 51. *Microsporum cookei.* The macroconidium is ellipsoid and rough walled. Bar is 10 μm.

Microsporum distortum di Menna et Marples, 1954

Description: Colonies fast growing, velvety to cottony, flat, often with radial grooves, white to tan or buff with a reverse yellow tan color. Macroconidia are 3- to 10-celled, thick-celled, roughened, irregular, contorted, and distorted. Microconidia are 1-celled and clavate to pyriform. *In vitro* hair perforation test is positive.

Salient characteristics: *Microsporum distortum* colonies are fast growing, velvety to cottony, white to tan, and produce highly distorted, thick-walled macroconidia. *Microsporum distortum* differs from *M. audouinii* by perforating hair and growing profusely on polished rice grains.

80

Microsporum ferrugineum Ota, 1921

Description: Two colony types are prevalent. Colonies of the first type are slow growing, glabrous, heaped, wrinkled, often with furrows and folds, and yellow to rust in color, often with a dull orange reverse. Colonies of the second type are slow growing, flat, spreading, leathery to downy, and white in color. Mycelium is usually sterile and characterized by being irregularly branched. Long, straight, thick-walled, heavily segmented hyphae are often present. *In vitro* hair perforation test is negative.

Salient characteristics: *Microsporum ferrugineum* forms slow-growing, flat to heaped, wrinkled, white or yellow to rust colonies consisting of a sterile mycelium, which may be branched and is often hypersegmented. Fusiform macroconidia typically are produced on rice grains or dilute Sabouraud dextrose agar.

Microsporum gallinae (Mégnin) Grigorakis, 1929

Description: Colonies are moderately fast growing, velvety to woolly or cottony, more or less wrinkled, white to gray, turning pink to buff with age, producing a strawberry-red diffusible pigment. Macroconidia are 2- to 10-celled (usually 5 to 6) smooth to slightly echinulate, clavate to cigar shaped, with thin cell walls. Microconidia are 1-celled and ovoid to pyriform. *In vitro* hair perforation test is negative.

Salient characteristics: *Microsporum gallinae* forms velvety to cottony, white to pink colonies that produce a diffusible strawberry-red pigment, and clavate to cigar-shaped macroconidia that are often slightly curved with fine echinulations at their apices. *Microsporum gallinae* differs from *Trichophyton megninii* by not requiring L-histidine for growth and by producing echinulate macroconidia.

Figure 52. *Microsporum gypseum*. The thin septal and outer cell walls of the macroconidia are the same thickness. Bar is 10 μm.

Microsporum gypseum (Bodin) Guiart et Grigorakis, 1928

Description: Colonies are rapid growing, granular, cinnamon tan in color with a buff to reddish brown reverse. Macroconidia are 3- to 9-celled, elliptic, with thin outer cell and septal walls, as well as an annular frill. There is a tendency for the outer cell wall to collapse slightly between each septum. Microconidia are 1-celled and clavate. *In vitro* hair perforation test is positive.

Salient characteristics: *Microsporum gypseum* forms buff to cinnamon tan colonies producing elliptic macroconidia having thin outer cell and septal walls, as well as a distinct annular frill.

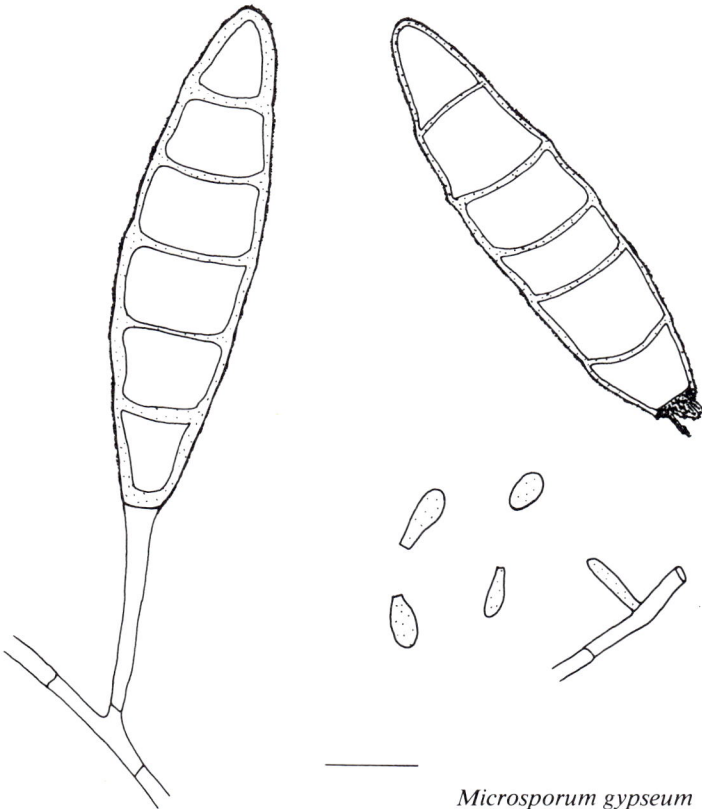

Microsporum gypseum

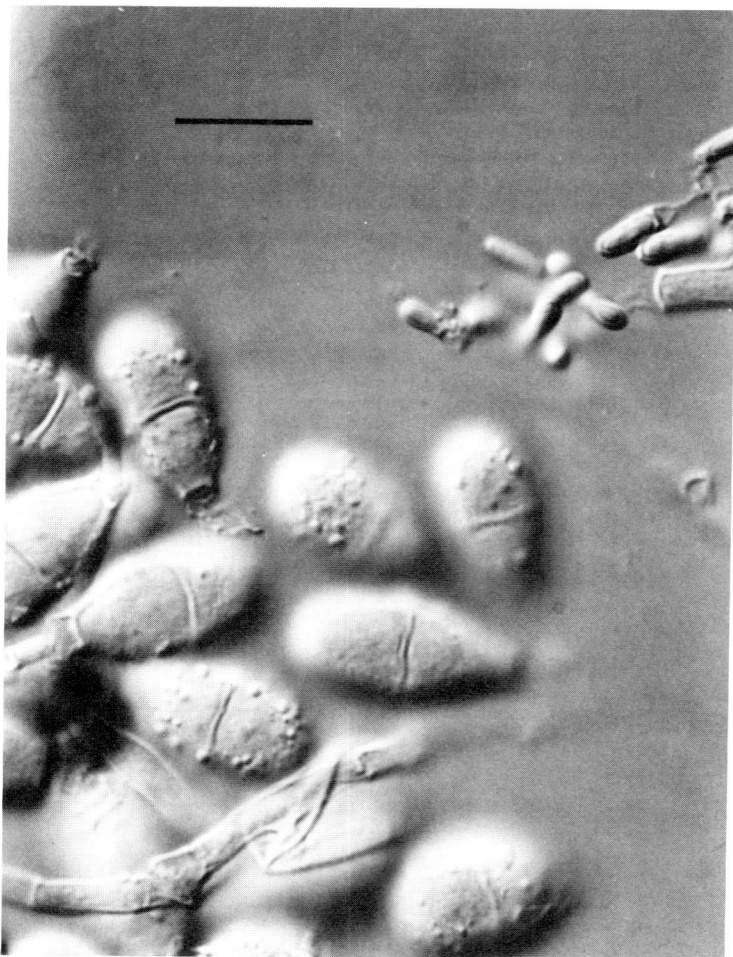

Figure 53. *Microsporum nanum.* The macroconidia are often 2-celled. Bar is 10 μm.

Microsporum nanum Fuentes, 1956

Description: Colonies are rapid growing, thin, spreading, powdery, velvety or flat, cottony, often with some radial, shallow furrows, dark buff with a reddish brown reverse color. Macroconidia are 1- to 4-celled (usually 2), rough, obovoid to pyriform, with thin walls. Microconidia are 1-celled and clavate. *In vitro* hair perforation test is positive.

Salient characteristics: *Microsporum nanum* forms powdery, buff colonies with a red brown reverse coloration and 2- to 3-celled macroconidia that are rough, obovoid to pyriform, and thin walled.

Microsporum vanbreuseghemii Georg, Ajello, Friedman et Brinkman, 1962

Description: Colonies are fast growing, flat, spreading, granular to velvety, cream yellow to lavender pink with a lemon yellow diffusible pigment and a yellow to orange reverse. Macroconidia are 6- to 13-celled, cylindrical to fusiform, rough, thick walled. Microconidia are 1-celled and obovoid to pyriform. *In vitro* hair perforation test is positive.

Salient characteristics: *Microsporum vanbreuseghemii* forms granular to velvety, flat, spreading, yellow to lavender pink colonies having thick-walled pencil-shaped macroconidia.

83

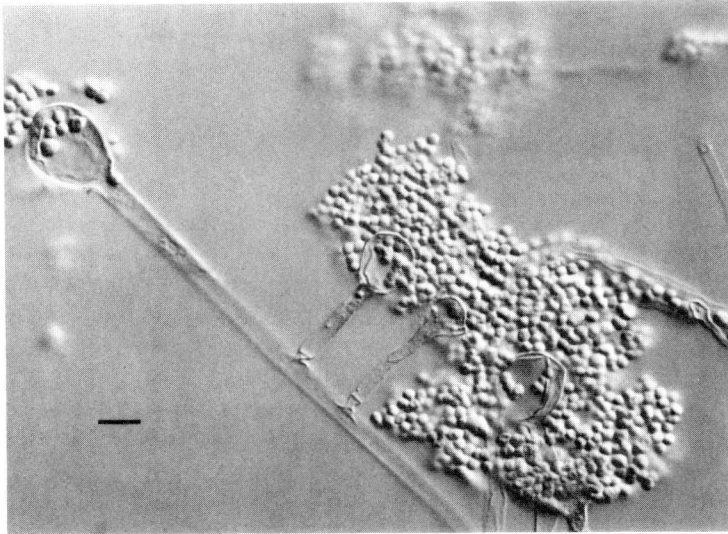

Figure 54. *Mucor miehei.* Sporangiospores are around the columellae. Bar is 10 μm.

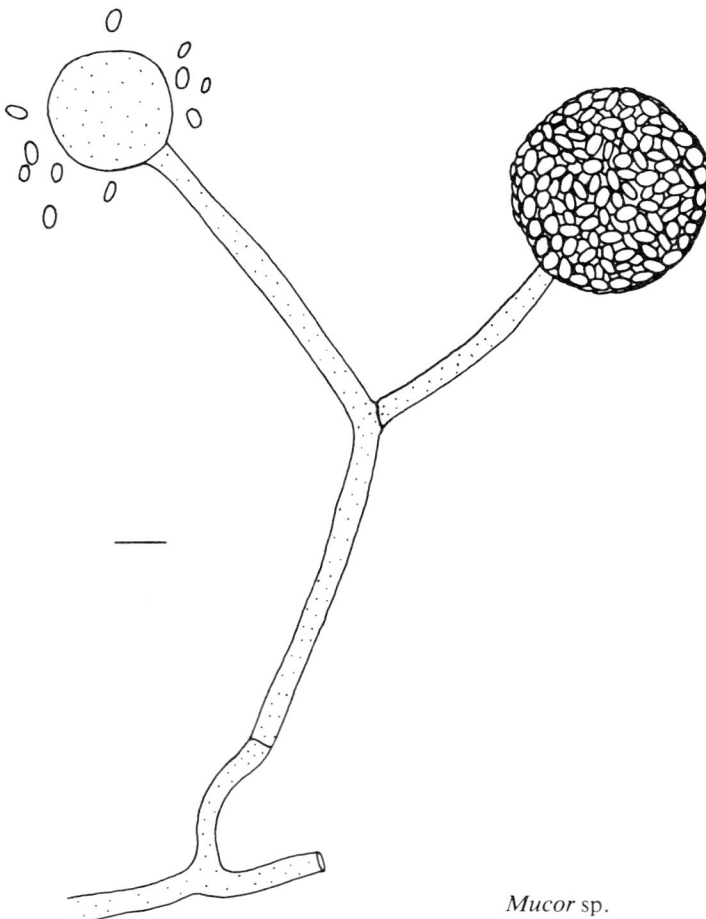

Mucor sp.

Mucor Micheli ex Saint-Amans, 1821

Description: Colonies are rapid growing, cottony, at first white, becoming olive gray to brown with a white reverse color. Sporangiophores are erect, solitary, branched or simple, arising from the hyphae. Rhizoids and stolons are absent. Sporangia are globose with dissolving walls, which often leave a basal collarette. Columellae are variable, hyaline to dematiaceous. Sporangiospores are 1-celled, globose to ellipsoid, smooth. If zygospores are present, they arise from the mycelium.

Salient characteristics: Members of the genus *Mucor* produce neither rhizoids nor stolons. They form erect sporangiophores bearing columellae and globose sporangia. *Mucor* differs from *Rhizopus* by having branching sporangiophores and by the absence of stolons and rhizoids; from *Rhizomucor* by the absence of stolons and rhizoids; and from *Absidia* by having globose sporangia without a swelling in the apex of the sporangiophore where it merges with the sporangium.

Laboratory precautions: Handle with care, but special precautions are not necessary.

Key references:
Schipper, 1978a
Zycha, Siepmann, and Linnemann, 1969

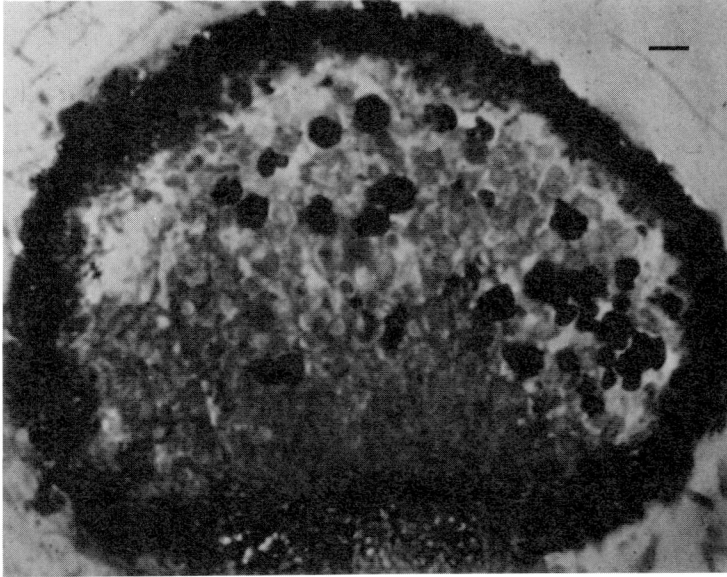

Figure 55. *Neotestudina rosatii.* The ascospores can be seen within this sectioned ascostroma. Bar is 10 μm. (Courtesy of L. Ajello.)

Neotestudina Segretain et Destombes, 1961

Description: Colonies are slow growing, woolly, compact, folded, gray to brown with a dark brown reverse color. Ascostromata are without ostioles and are black, carbonaceous, and globose to ellipsoidal. Asci are in the center of the ascostromata, they are globose, or subglobose to clavate, thick walled, bitunicate, and contain 8 ascospores. Ascospores are 2-celled with a single transverse septum that is sharply constricted, ellipsoid, variable in size, dark brown, and smooth.

Salient characteristics: *Neotestudina* is characterized by producing ascostromata composed of plates of radiating cells, globose to subglobose asci, and smooth, evenly pigmented ascospores with a germ pore at each end.

Laboratory precautions: Handle with care, but special precautions are not necessary.

Key reference:
Hawksworth, 1979

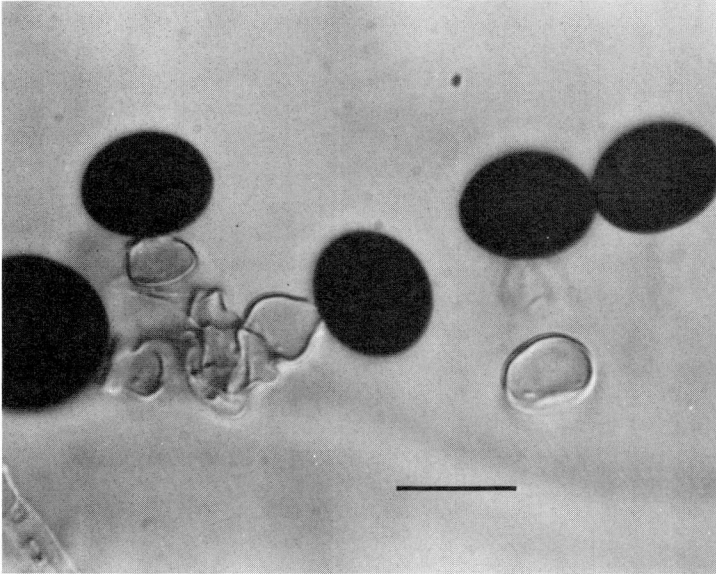

Figure 56. *Nigrospora oryzae.* The black, horizontally flattened conidia develop on hyaline conidiophores that are centrally swollen. Bar is 10 μm.

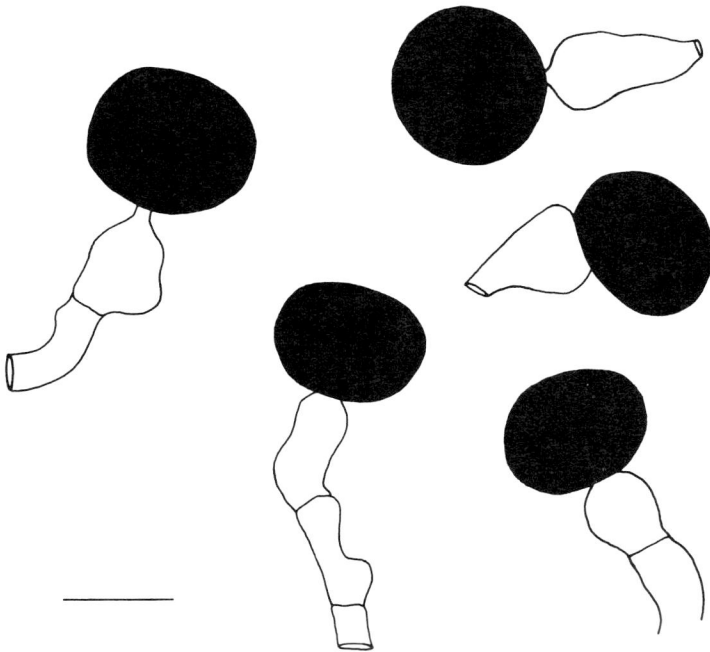

Nigrospora oryzae

Nigrospora Zimmerman, 1902

Description: Colonies are rapid growing, compact, woolly, at first white, becoming gray with black areas and a black reverse color. Conidiophores are hyaline, swollen in the middle, and abruptly taper to the point at which the conidia are formed. Conidia are 1-celled, subglobose to ovoid, horizontally flattened, solitary, smooth, and black.

Salient characteristics: Isolates of *Nigrospora* form black, 1-celled, horizontally flattened conidia arising solitarily from hyaline, centrally swollen conidiophores.

Laboratory precautions: Handle with care, but special precautions are not necessary.

Key reference:
McGinnis, 1980

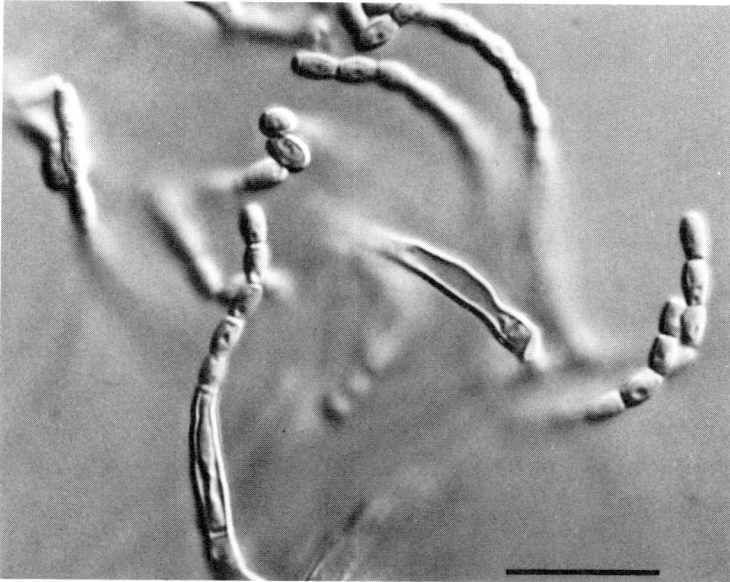

Figure 57. *Paecilomyces variotii.* The conidia form entangled chains. Bar is 10 μm.

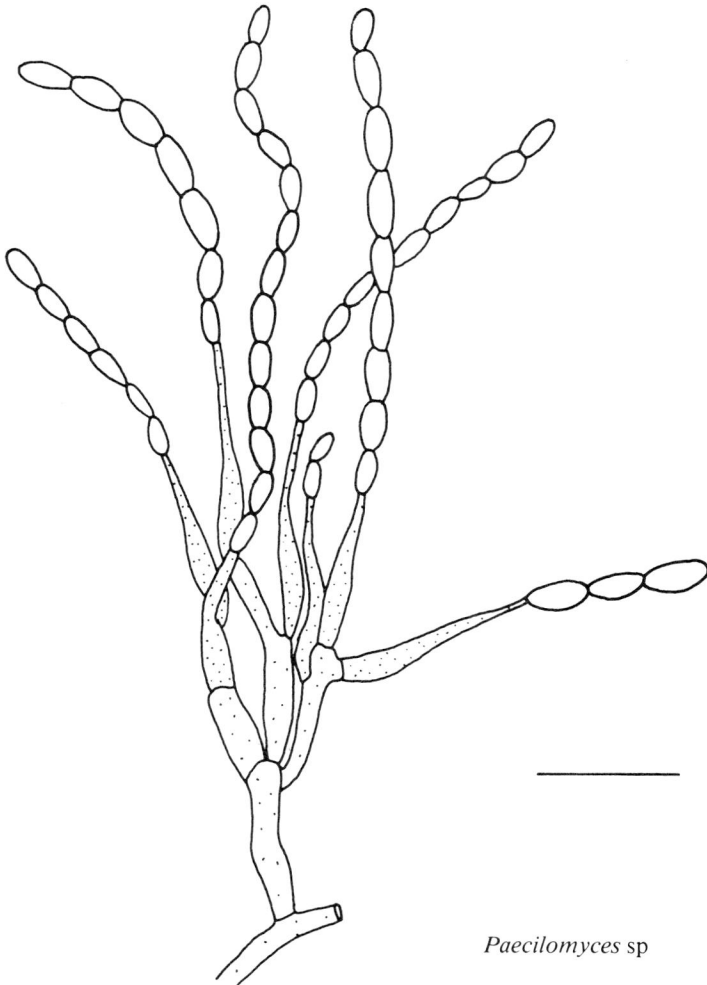

Paecilomyces sp

Paecilomyces Bainier, 1907

Description: Colonies are rapid growing, flat, cottony to ropy in texture, initially white, becoming yellow, yellow brown to olive brown, with a dirty white, buff, or brown reverse color. A sweet aromatic odor may be associated with older cultures. Erect conidiophores branch at their apices, which bear phialides. Phialides are swollen at their bases, taper toward their apices, are usually in pairs, groups, or verticils, and frequently form a penicillus. Conidia are 1-celled, hyaline to darkly colored, smooth or rough, ovoid to fusoid, and form entangled basipetal chains.

Salient characteristics: *Paecilomyces* species form entangled chains of 1-celled conidia from phialides that are basally swollen, tapering toward their apices, which are formed on branched conidiophores. *Paecilomyces* differs from *Penicillium* by having basally swollen phialides that taper toward their apices.

Laboratory precautions: Handle with care, but special precautions are not necessary.

Key references:
Brown and Smith, 1957
Samson, 1974

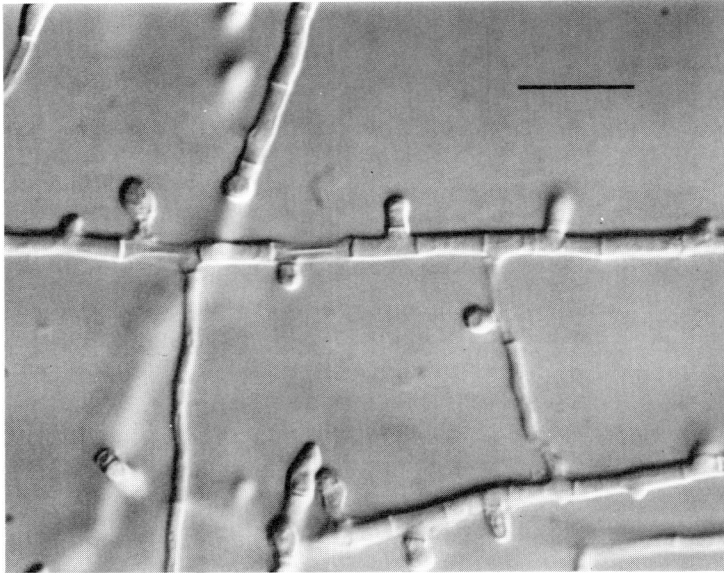

Figure 58. *Paracoccidioides brasiliensis.* 1-celled conidia are occasionally produced by some isolates, 25°C. Bar is 10 μm.

Paracoccidioides
de Almeida, 1930

Description: Colonies are slow growing, leathery, flat to wrinkled, woolly, cottony, or velvety, white cream tan or brown, with a yellowish brown to brown reverse color. Conidia are 1-celled, truncate, with a broad base and rounded apex, cylindrical, solitary, and hyaline; they are rarely produced. At 37°C, *Paracoccidioides* produces a yeast form having multiple blastoconidia. Yeast cells are globose to subglobose, blastoconidia are attached to parent cell by narrow neck.

Salient characteristics: *Paracoccidioides brasiliensis* forms slow-growing, sterile colonies with occasional conidia in some isolates. A multiple-budding yeast is formed at 37°C.

Laboratory precautions: Handle with caution in a biological safety cabinet.

Key reference:
McGinnis, 1980

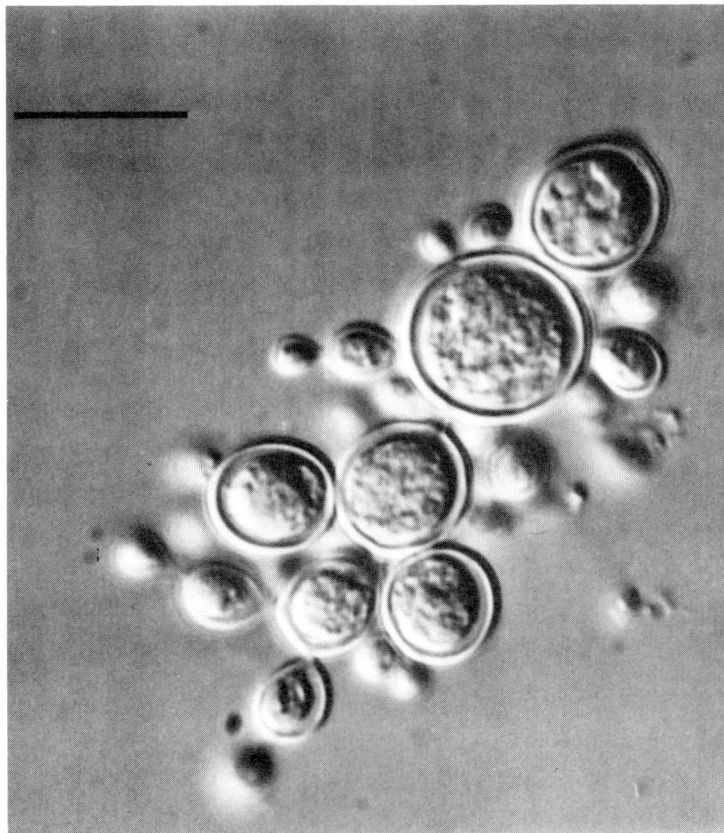

Figure 59. *Paracoccidioides brasiliensis.* The yeast cells show multiple budding, 37°C. Bar is 10 μm.

88

Figure 60. *Penicillium* sp. The conidial head resembles a brush. Bar is 10 μm. (Reproduced by permission of Academic Press from M. McGinnis. *Laboratory Handbook of Medical Mycology,* 1980.)

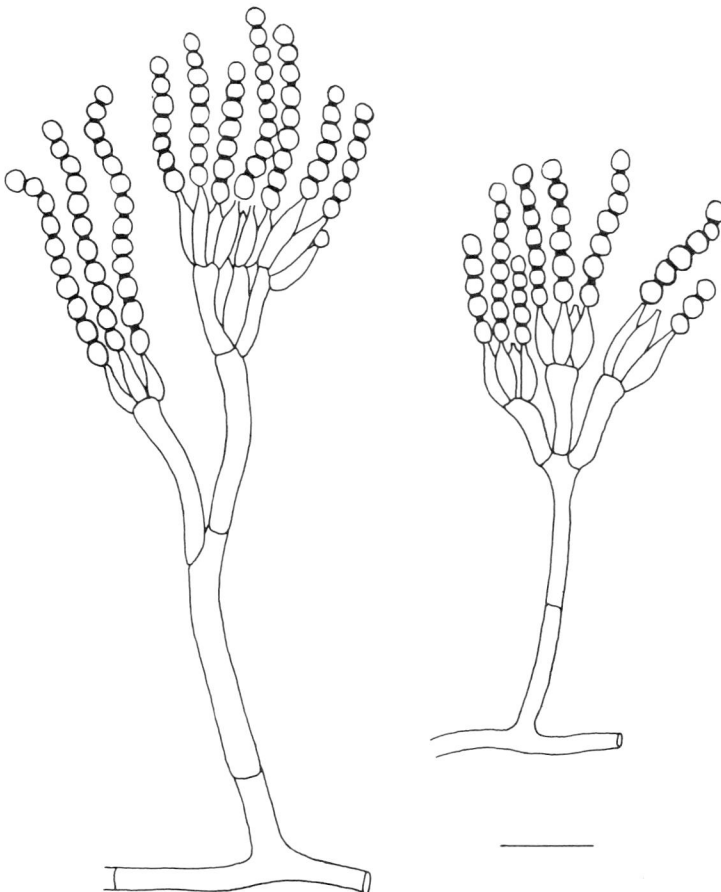

Penicillium sp.

Penicillium Link ex Gray, 1821

Description: Colonies are rapid growing, flat, velvety, woolly, or cottony, initially white, becoming blue green, gray green, olive gray, or some similar color. Drops of fluid, sclerotia, cleistothecia, or any combination of these may be found. Conidiophores are erect, branched, hyaline or pigmented. Phialides are flask shaped, arising in groups at the apices of terminal branches (metulae) of the conidiophore. Conidia are 1-celled, globose to ovoid, smooth or rough, hyaline or pigmented, arising as basipetal chains. The entire conidial structure resembles a brush.

Salient characteristics: The genus *Penicillium* is characterized by rapidly growing colonies having conidial structures resembling brushes. *Penicillium* differs from *Paecilomyces* by having flask-shaped phialides and globose to subglobose conidia; from *Gliocladium* by having chains of conidia; and from *Scopulariopsis* by forming phialides.

Laboratory precautions: Handle with care, but special precautions are not necessary.

Key references:
Raper and Thom, 1949
Samson, Hadlok, and Stolk, 1977
Samson, Stolk, and Hadlok, 1976

Phaeococcomyces de Hoog, 1979

Description: Colonies are slow growing, restricted, slimy, yeast-like to wrinkled, and black in color. Yeast cells produce 1-celled, globose to ellipsoid conidia that are nearly hyaline to dark brown to black; pseudohyphae may be present; hyphae are absent.

Salient characteristics: *Phaeococcomyces* is a genus that includes black yeasts that do not form hyphae. It is usually one of the forms associated with such fungi as *Exophiala jeanselmei* and *Wangiella dermatitidis*. Due to a problem of nomenclature, the genus *Phaeococcomyces* has been established as a replacement for *Phaeococcus*.

Laboratory precautions: Handle with care, but special precautions are not necessary.

Key reference:
de Hoog, 1977

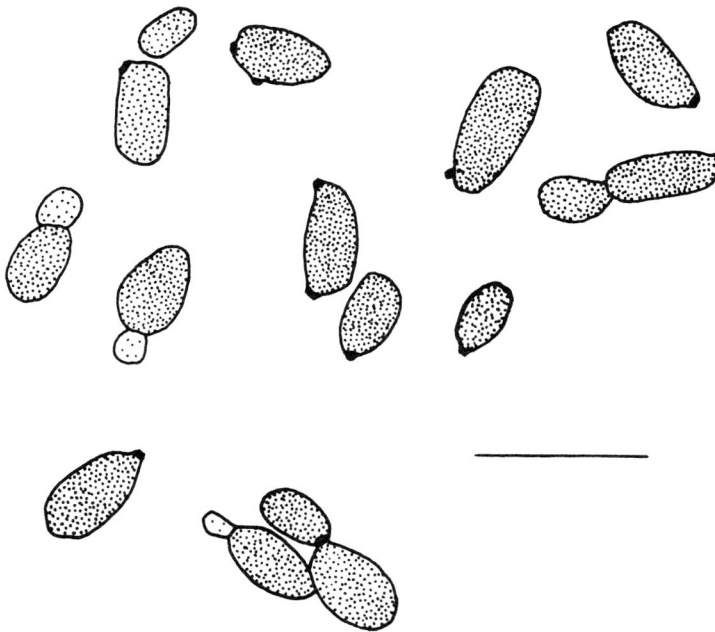

Phaeococcomyces exophialae

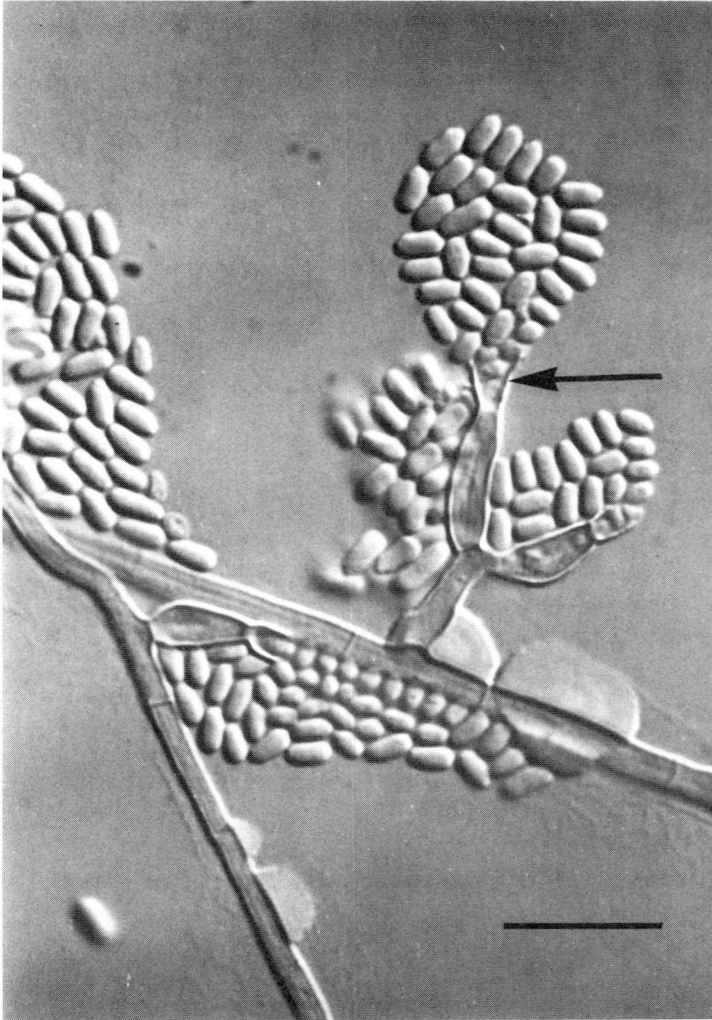

Figure 61. *Phialophora verrucosa.* The phialides have cuplike collarettes (arrow). Bar is 10 μm.

Phialophora Medlar, 1915

Description: Colonies are moderately rapid growing, woolly to cottony, gray, gray brown to almost black, with a reverse color of iron gray to black. Conidiophores when present are short and hyphalike. Phialides are cylindrical to flask shaped, with collarettes, solitary or in clusters, along hyphae or on conidiophores, dematiaceous. Conidia are 1-celled, hyaline to brown, smooth, ovoid to cylindrical, occurring as a ball at the apex of the phialide.

Salient characteristics: Balls of 1-celled conidia accumulating at the apices of dematiaceous phialides that are usually flask shaped with a distinct collarette characterize the more common species of *Phialophora*. *Phialophora* differs from *Exophiala* by having phialides; it differs from *Wangiella* by having phialides with collarettes.

Laboratory precautions: Handle with care, but special precautions are not necessary.

Key reference:
McGinnis, 1978

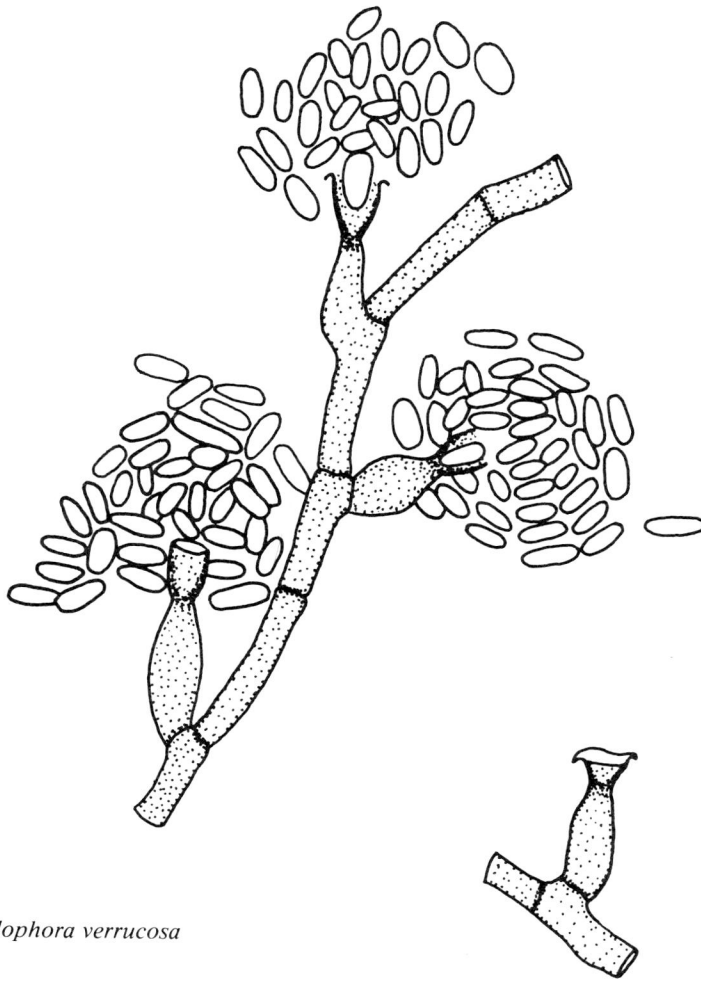

Phialophora verrucosa

Key to the Human Pathogenic Species of *Phialophora*

1.   Phialides with saucer-shaped or flared collarettes .... *P. richardsiae*

1.´   Phialides without saucer-shaped or flared collarettes 2

    2.   Collarettes vase shaped, darkly pigmented;
          phialides flask shaped .... *P. verrucosa*

    2´   Collarettes not vase shaped, but with parallel walls.... 3

3.   Colonies yeastlike, cottony, cream, pink to gray;
    phialides not separated from hyphae by a septum; conidia 1-celled, often curved .... *P. hoffmannii*

3.´   Colonies not yeastlike, gray.................................. 4

    4.   Phialides cylindrical to obclavate, elongate, hyaline
          to pale brown .... *P. parasitica*

    4.´   Phialides cylindrical to lageniform, short, hyaline
          to pale brown .... *P. repens*

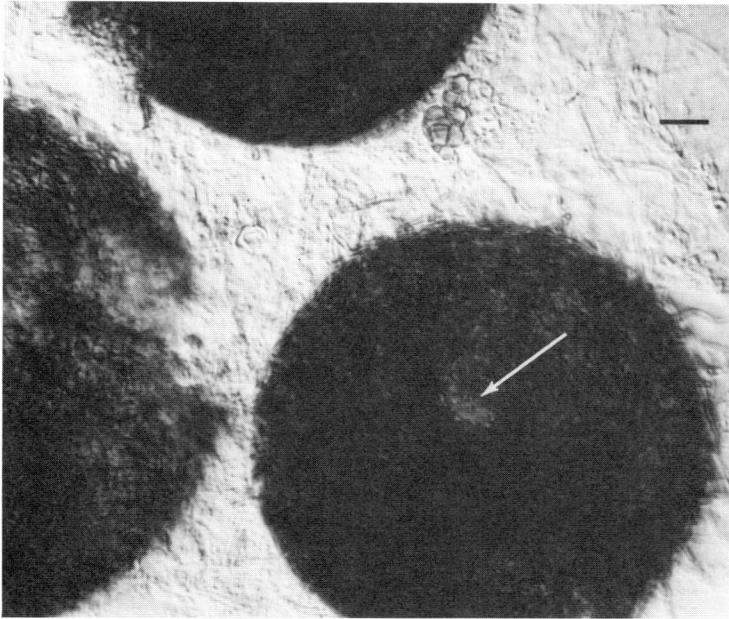

Figure 62. *Phoma* sp. The pycnidium has an ostiole (arrow). Bar is 10 μm.

Phoma Saccardo, 1880 *nom. cons.*

Description: Colonies are moderately rapid growing, flat, spreading, woolly to cottony, often largely submerged in the medium, at first white, becoming olivaceous green to olivaceous gray, with a dark reverse color. There is a reddish purple to yellowish brown diffusible pigment in some species. Pycnidia are subglobose to pyriform, with one to several ostioles, with or without a short neck, membranous, dark brown to black. Phialides arise from the inner lining of the pycnidia. Conidia are 1-celled, globose to oblong with rounded ends, hyaline, and typically with two oil drops.

Salient characteristics: *Phoma* species form dark, ostiolate pycnidia that produce phialides and 1-celled, hyaline, globose to oblong conidia that have two oil droplets in each conidium. *Phoma* differs from *Pyrenochaeta* by the absence of setae.

Laboratory precautions: Handle with care, but special precautions are not necessary.

Key reference:
Sutton, 1980

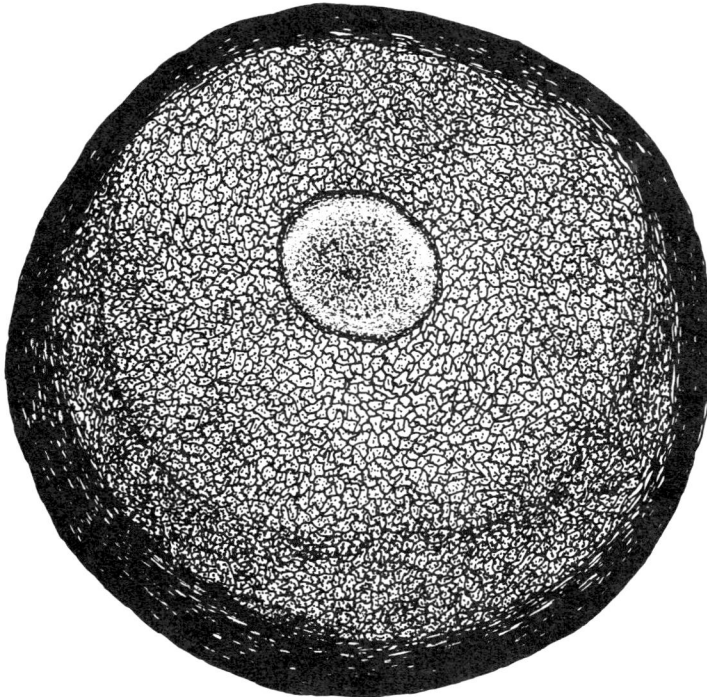

Phoma sp.

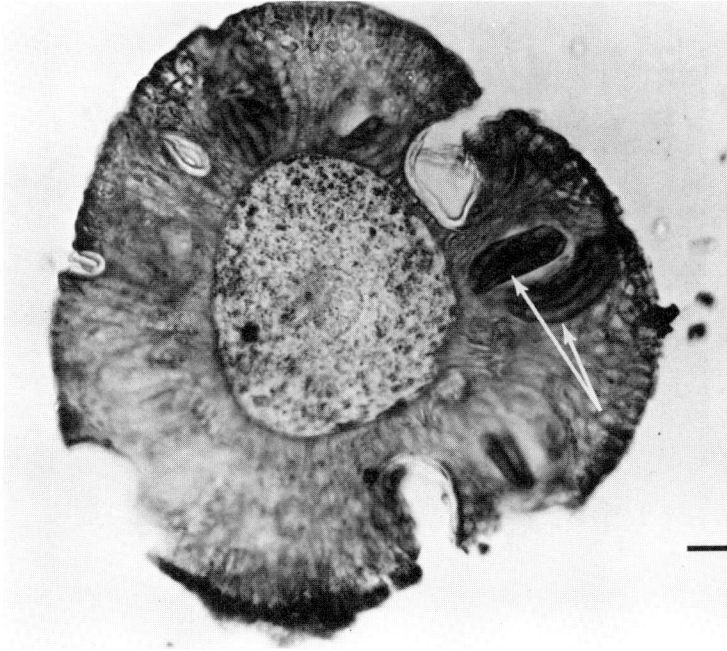

Figure 63. *Piedraia hortae.* The ascostroma is surrounding a hair. The asci (arrow) form within a locule. Bar is 10 μm. (Reproduced by permission from *Mycopathologia Mycologia et Applicata* 45:269–83, 1971.)

Piedraia Fonseca et Arêa Leão, 1928

Description: Colonies are slow growing, small, folded, velvety and dark brown to black in color. Ascostromata are subglobose to irregular in shape and are black. Asci are ellipsoid, solitary or in clusters, with 8 ascospores and ascus walls that readily dissolve. Ascospores are 1-celled, fusoid, curved, and tapering toward both ends to form whiplike extensions; they are hyaline to darkly pigmented.

Salient characteristics: *Piedraia* forms 1-celled, curved, tapering ascospores with an appendage at each end.

Laboratory precautions: Handle with care, but special precautions are not necessary.

Key reference:
Takashio and Vanbreuseghem, 1971

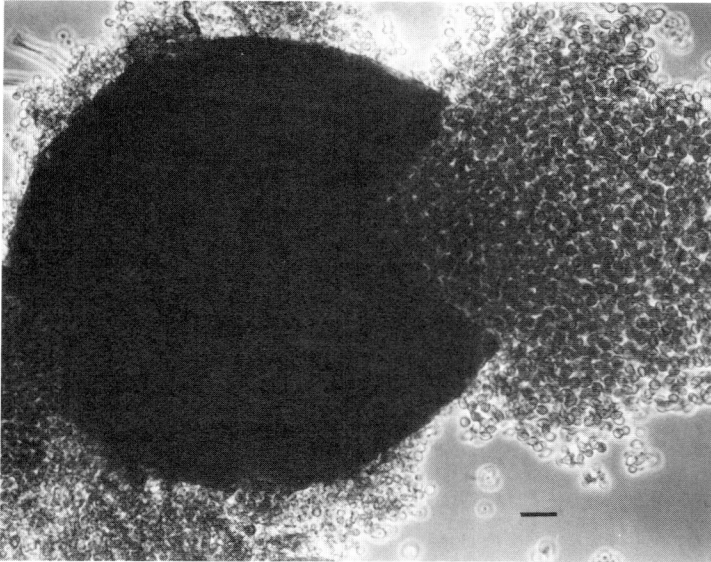

Figure 64. *Pseudallescheria boydii.* Ascospores are escaping from the cleistothecium. Bar is 10 μm.

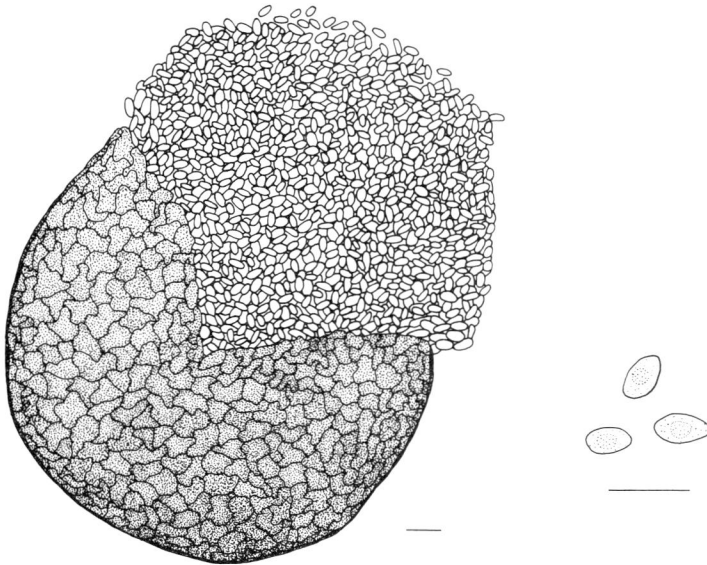

Pseudallescheria boydii

Pseudallescheria Negroni et Fischer, 1943

Description: Colonies are rapid growing, spreading, cottony, at first white, becoming pale smoky brown in color. Cleistothecia form just beneath the agar surface and are globose, light brown to black, without appendages and without ostioles. Asci are subglobose to globose with 8 ascospores; ascus walls readily dissolve to release ascospores. Ascospores are 1-celled, ovoid to ellipsoid, smooth, pale yellow brown to copper in color. *Graphium, Scedosporium,* or both forms may be present in the same isolate.

Salient characteristics: Globose, black cleistothecia forming beneath the agar and having 1-celled, ovoid, yellow brown to copper-colored ascospores are typical of the genus *Pseudallescheria. Pseudallescheria* differs from *Petriella* by forming nonstiolate cleistothecia. *Petriellidium* is a later synonym of *Pseudallescheria.*

Laboratory precautions: Handle with care, but special precautions are not necessary.

Key reference:
McGinnis, Padhye, and Ajello, 1981

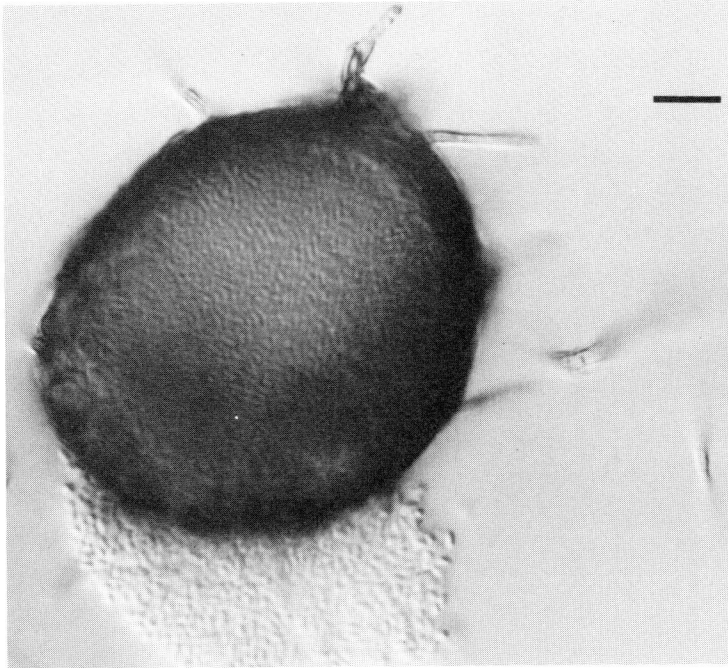

Figure 65. *Pyrenochaeta romeroi.* The pycnidium has setae. Bar is 10 μm.

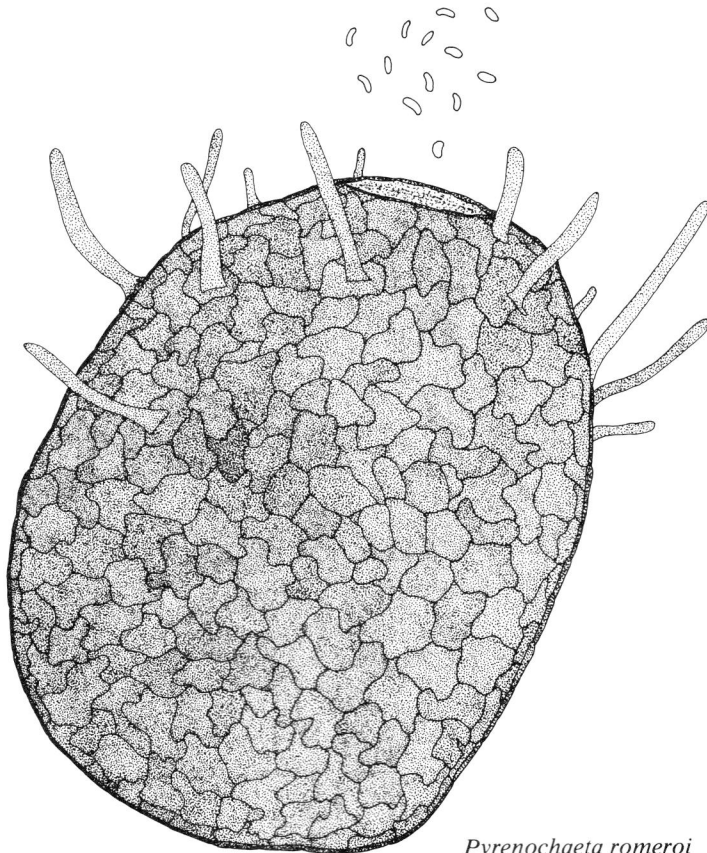

Pyrenochaeta romeroi

Pyrenochaeta de Notaris, 1849

Description: Colonies are moderately rapid growing, flat, woolly to cottony, at first white, becoming olivaceous green to olivaceous gray in color with a dark reverse. Pycnidia are globose to flask shaped, ostiolate, membranous to carbonaceous, brown to black, with setae arising from the upper portion of the pycnidia. Phialides arise from the inner lining of the pycnidia. Conidia are 1-celled, oval to cylindrical, hyaline, and may be slightly curved.

Salient characteristics: *Pyrenochaeta* species form dark, ostiolate pycnidia having black setae arising from the upper portion of the pycnidia, and form 1-celled, oval to cylindrical conidia. *Pyrenochaeta* differs from *Phoma* by having setae.

Laboratory precautions: Handle with care, but special precautions are not necessary.

Key reference:
Sutton, 1980

96

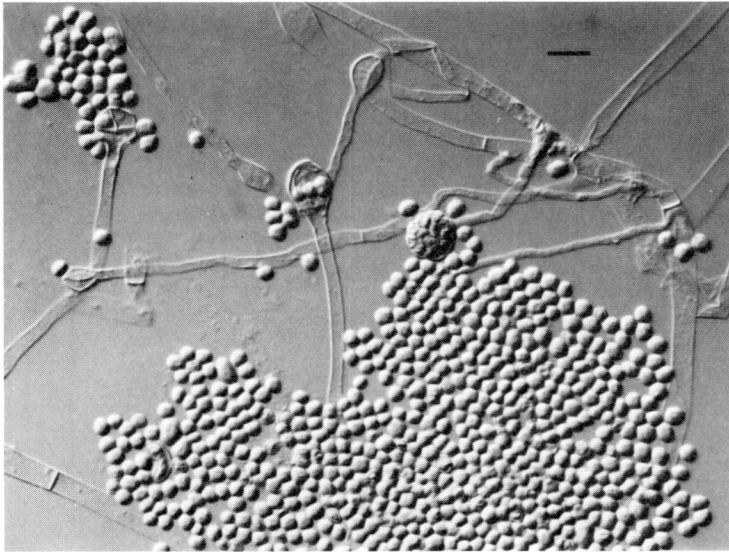

Figure 66. *Rhizomucor pusillus.* Sporangia and sporangiospores. Bar is 10 μm.

Rhizomucor (Lucet et Costantin) Wehmer ex Vuillemin, 1931

Description: Colonies are rapid growing, cottony, at first white, becoming smoky gray to brown. Growth is present at 50–55°C. Sporangiophores arise from stolons or aerial hyphae, are repeatedly branched, and gray. Rhizoids (usually poorly developed) are present. The sporangia are globose with columellae. Sporangiospores are 1-celled and globose. If zygospores are present, they are formed in the aerial hyphae.

Salient characteristics: Species of *Rhizomucor* form stolons, primitive rhizoids, branching sporangiophores, and have the ability to grow at a temperature of 50–55°C. *Rhizomucor* differs from *Mucor* by growing at 50–55°C and by having rhizoids and stolons; from *Rhizopus* by having branched sporangiophores and rhizoids not arising opposite the sporangiophores; and from *Absidia* by having globose sporangia and sporangiophores that are not swollen where they merge with the columellae.

Laboratory precautions: Handle with care, but special precautions are not necessary.

Key reference:
Schipper, 1978b

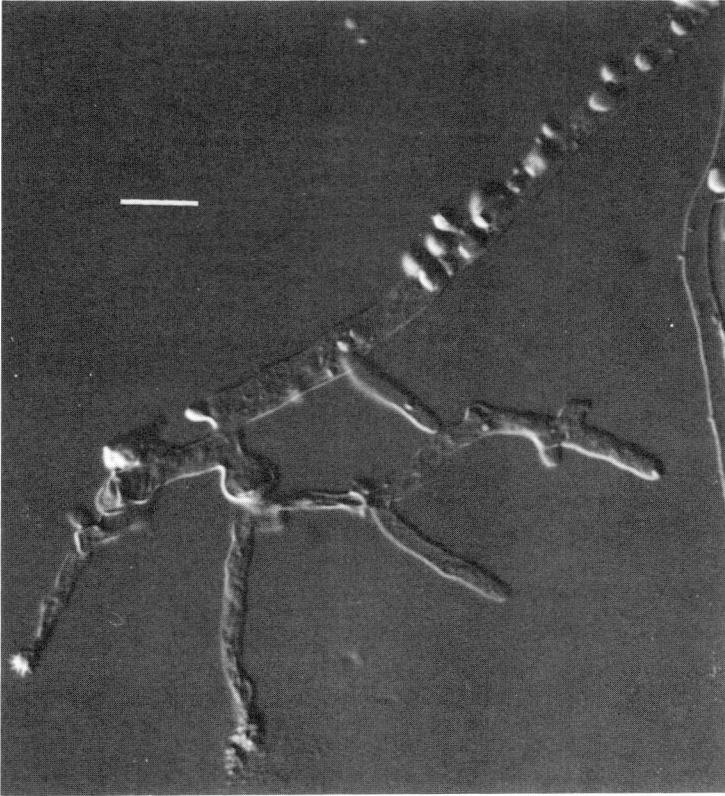

Figure 67. *Rhizomucor pusillus.* Poorly developed rhizoids are form-ed by this genus. Bar is 10 μm. (Reproduced by permission of Academic Press from M. McGinnis. *Laboratory Handbook of Medical Mycology,* 1980.)

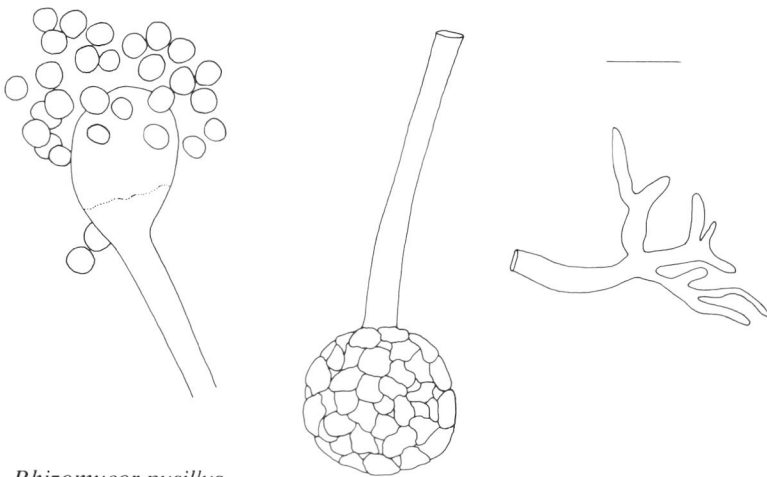

Rhizomucor pusillus

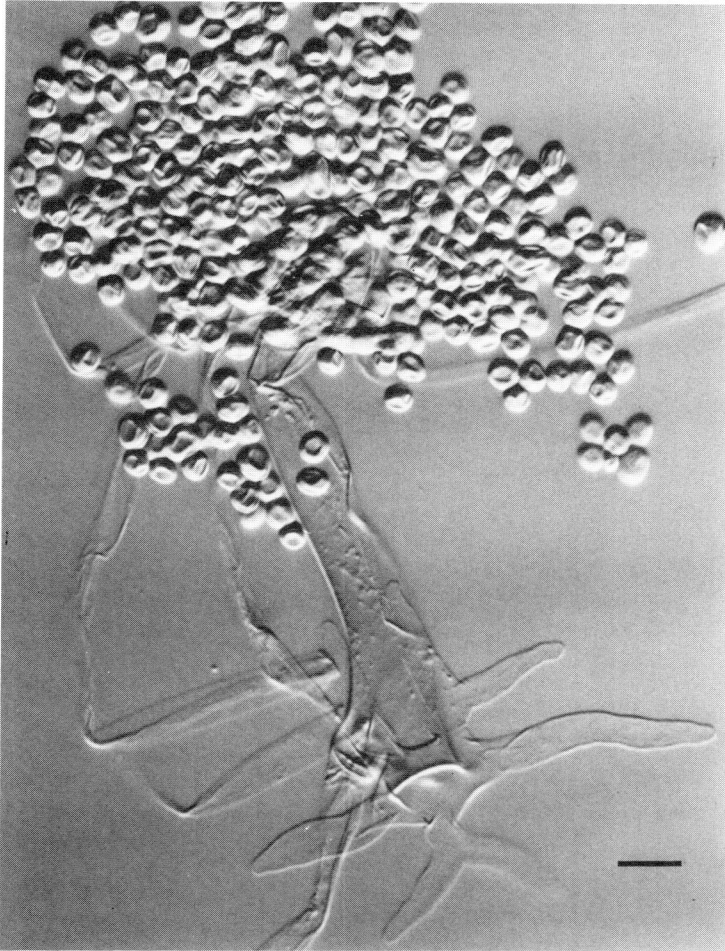

Figure 68. *Rhizopus oligosporus.* The rhizoids are arising opposite the sporangiophore. Bar is 10 μm.

Rhizopus Ehrenberg ex Corda, 1838

Description: Colonies are rapid growing, often filling the Petri dish, cottony and dense, at first white, becoming brownish gray in color. Sporangiophores are dark, solitary or in clusters, arising opposite rhizoids at a node. Sporangia are globose with flattened bases. Columellae are hemispherical. Sporangiospores are 1-celled, globose to ovoid, hyaline to brown, smooth or striated.

Salient characteristics: *Rhizopus* species form globose sporangia on simple sporangiophores that arise opposite rhizoids at the nodes of the stolons. *Rhizopus* differs from *Absidia, Mucor,* and *Rhizomucor* by having simple sporangiophores arising opposite rhizoids at nodes.

Laboratory precautions: Handle with care, but special precautions are not necessary.

Key reference:
Inui, Takeda, and Iizuka, 1965

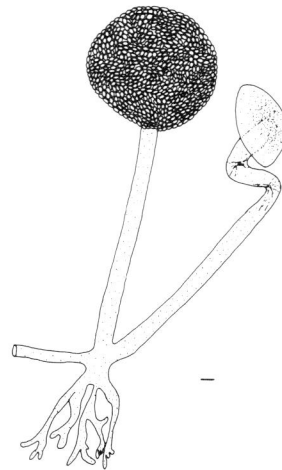

Rhizopus sp.

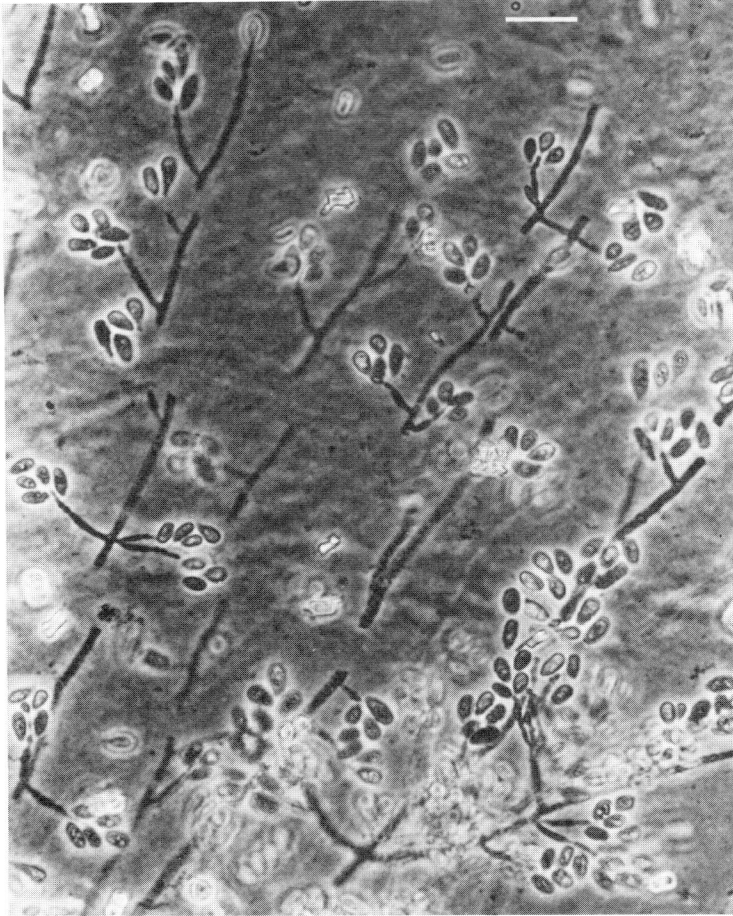

Figure 69. *Scedosporium apiospermum.* The conidia tend to accumulate at the apices of the annellides. Bar is 20 μm.

Scedosporium Saccardo ex Castellani et Chalmers, 1919

Description: Colonies are rapid growing, spreading, cottony, at first white, becoming pale smoky brown. Conidiophores are short or long, hyphalike. Conidiogenous cells are cylindrical annellides. Conidia are 1-celled, solitary or in balls, subglobose to elongate, smooth, pale brown. *Graphium, Pseudallescheria,* or other forms may be present also.

Salient characteristics: *Scedosporium* isolates form 1-celled, brownish conidia that arise solitarily on hyphalike conidiophores, in clusters at the apices of annellides, or in both ways. *Scedosporium apiospermum,* previously known as *Monosporium apiospermum,* is presently the only known pathogen in the genus.

Laboratory precautions: Handle with care, but special precautions are not necessary.

Key reference:
McGinnis, 1980

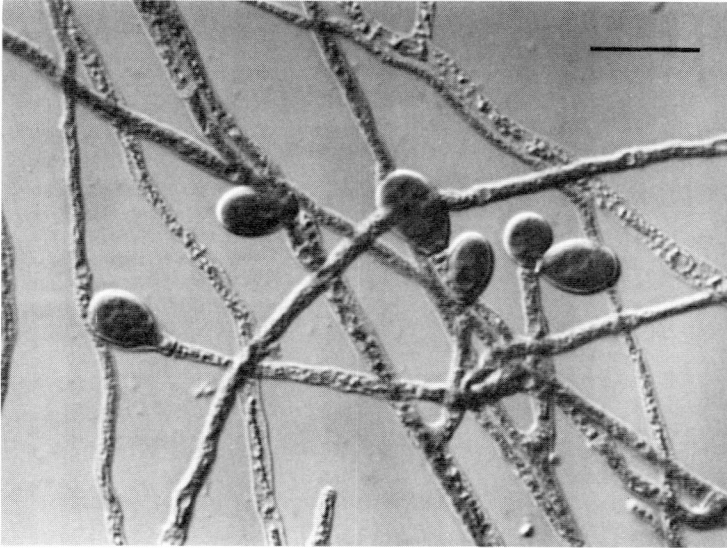

Figure 70. *Scedosporium apiospermum.* The conidia may be solitary or in balls. Bar is 10 μm.

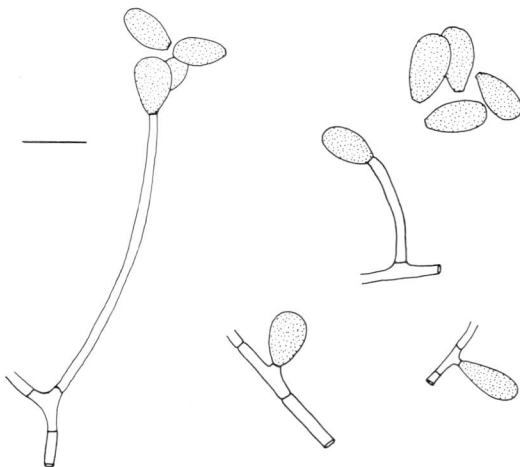

Scedosporium apiospermum

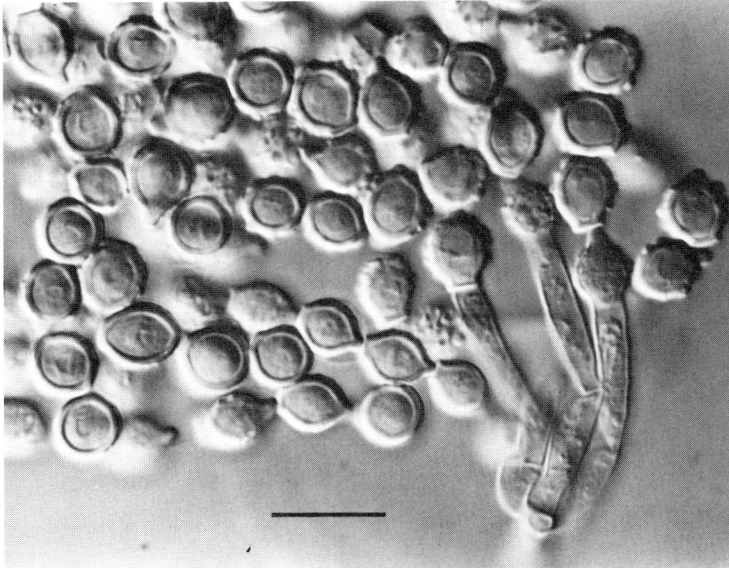

Figure 71. *Scopulariopsis brevicaulis.* The rough-walled conidia arise in chains from the annellides. Bar is 10 μm.

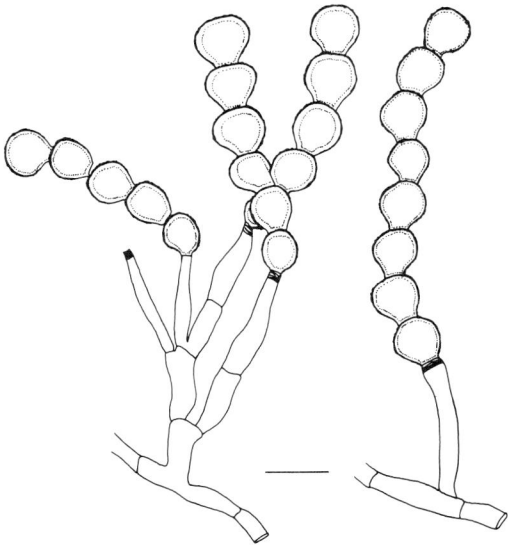

Scopulariopsis brevicaulis

Scopulariopsis Bainier, 1907

Description: Colonies are moderately rapid growing, granular to powdery, initially white, becoming light brown or buff tan, reverse color usually is tan. Conidiophores are hyphalike and simple or branched. Annellides are solitary, in clusters, or form a penicillus; they are cylindrical and slightly swollen. Conidia are 1-celled, globose to pyriform, smooth, but more commonly rough walled, truncate, forming basipetal chains. A *Trichurus* form (synnemata) may be present.

Salient characteristics: Isolates of *Scopulariopsis* are granular to powdery, light brown to buff tan, and form cylindrical annellides and chains of 1-celled, rough-walled conidia having truncate bases. *Scopulariopsis* differs from *Penicillium* by forming annellides.

Laboratory precautions: Handle with care, but special precautions are not necessary.

Key reference:
Morton and Smith, 1963

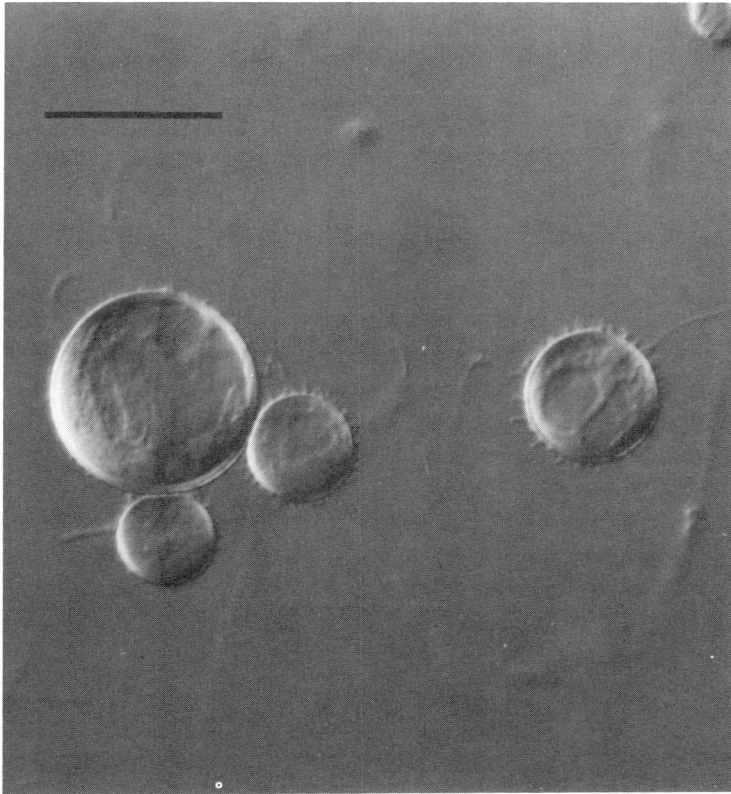

Figure 72. *Sepedonium* sp. The conidia are rough walled and resemble *Histoplasma capsulatum*. Bar is 10 μm.

Sepedonium Link ex Greville, 1924

Description: Colonies are rapid growing, woolly to cottony, white, cream, or tan in color. Conidiophores are hyphalike, short or long, hyaline, simple or branched. Conidia are 1-celled, terminal, solitary, globose to ovoid, large, roughened, thick walled, and hyaline to amber in color. A *Verticillium* stage may be present.

Salient characteristics: Most isolates of *Sepedonium* form 1-celled, thick-walled, roughened, amber conidia on hyphalike conidiophores. *Sepedonium* differs from *Histoplasma* by not being dimorphic; it differs from *Chrysosporium* by having large, thick-walled, roughened, amber conidia and usually an associated phialidic state.

Laboratory precautions: Handle with care, but special precautions are not necessary.

Key references:
Barron, 1968
Carmichael, 1962

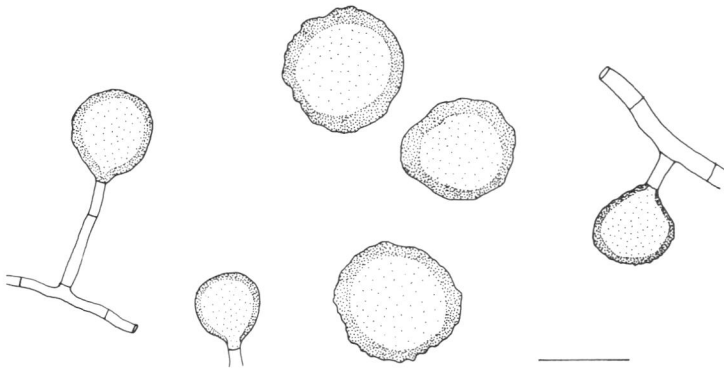

Sepedonium sp.

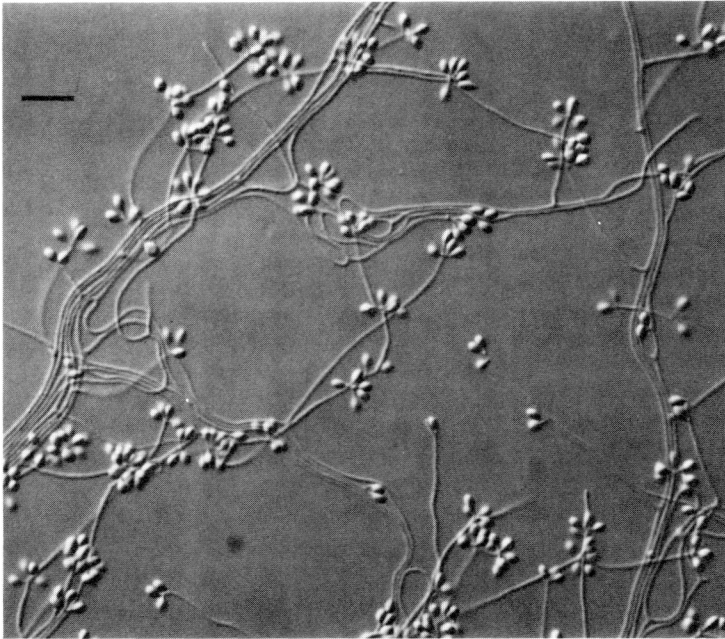

Figure 73. *Sporothrix schenckii.* The conidia develop upon short denticles from sympodial conidiophores, 25°C. Bar is 10 μm.

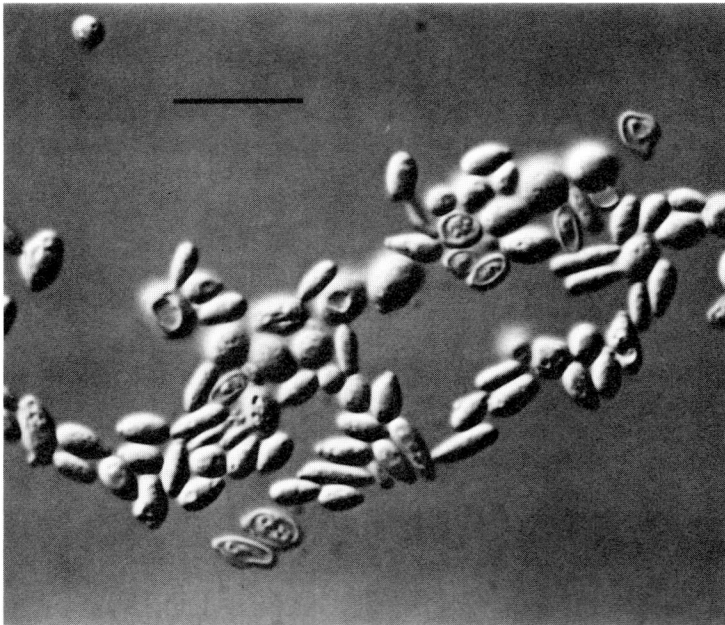

Figure 74. *Sporothrix schenckii.* The yeast form developed at 37°C on an enriched medium. Bar is 10 μm.

Sporothrix Hektoen et Perkins, 1900

Description: Colonies are rapid growing, moist, wrinkled, leathery to velvety in texture, at first white, becoming cream to dark brown or black at 25°C. Conidiophores typically are present and are hyphalike, hyaline, septate, sympodial, and often have an inflated apex. Conidia are of two kinds: the first kind are 1-celled, globose to clavate, arise solitarily on denticles, and often forming rosettes; the second kind are 1-celled, thick walled, dematiaceous, and arise along the hyphae. *Sporothrix schenckii* is dimorphic and develops a yeast form at 37°C.

Salient characteristics: *Sporothrix schenckii* is recognized by its dimorphic nature. At 25°C, it forms sympodial conidiophores having 1-celled hyaline conidia on denticles that frequently occur as rosettes at the apices of swollen condiophores, and dematiaceous 1-celled conidia along the hyphae. At 37°C, the fungus grows as a yeast.

Laboratory precautions: Handle with care, but special precautions are not necessary.

Key reference:
de Hoog, 1974

Sporothrix schenckii

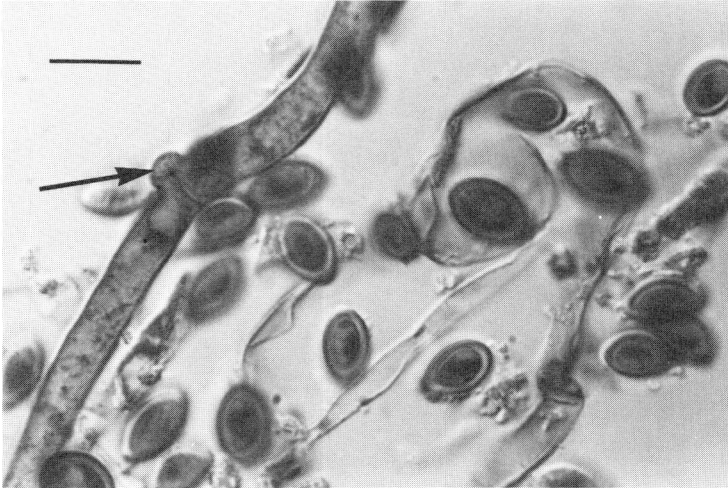

Figure 75. *Sporotrichum aureum.* The conidia are golden orange in color. A clamp connection can be seen at the arrow. Bar is 10 μm.

Sporotrichum Link ex Gray, 1821

Description: Colonies are moderately rapid growing, velvety to granular, at first white, becoming cinnamon buff to pinkish buff, or golden orange to ochraceous orange. Clamp connections are present at the septa. Conidia are 1-celled, truncate with broad bases, solitary, thick walled, golden yellow. An annular frill is formed when the conidia are released from their conidiophores.

Salient characteristics: Isolates of *Sporotrichum* form thick-walled, golden, 1-celled conidia with broad bases and a distinct annular frill. Clamp connections are present. *Sporotrichum* differs from *Chrysosporium* and *Sporothrix* by typically having clamp connections and thick walled, 1-celled, golden yellow conidia with annular frills. There are no human pathogens in the genus *Sporotrichum*.

Laboratory precautions: Handle with care, but special precautions are not necessary.

Key reference:
McGinnis, 1980

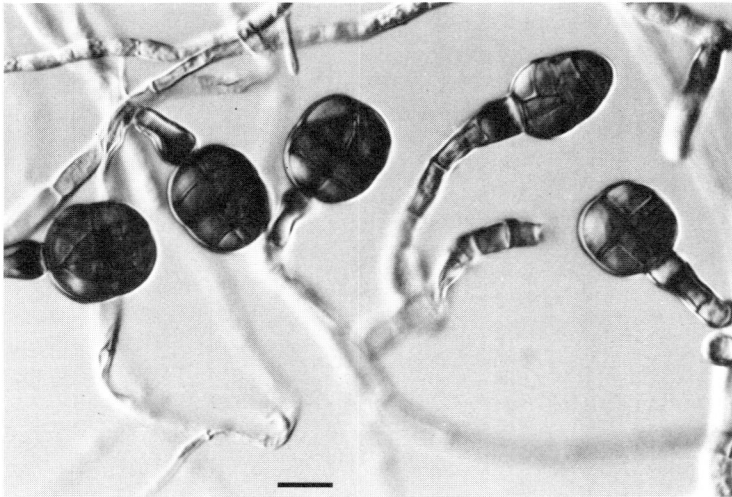

Figure 76. *Stemphylium sarcinaeforme.* The muriform conidia develop at the apices of percurrent conidiophores. Bar is 10 μm.

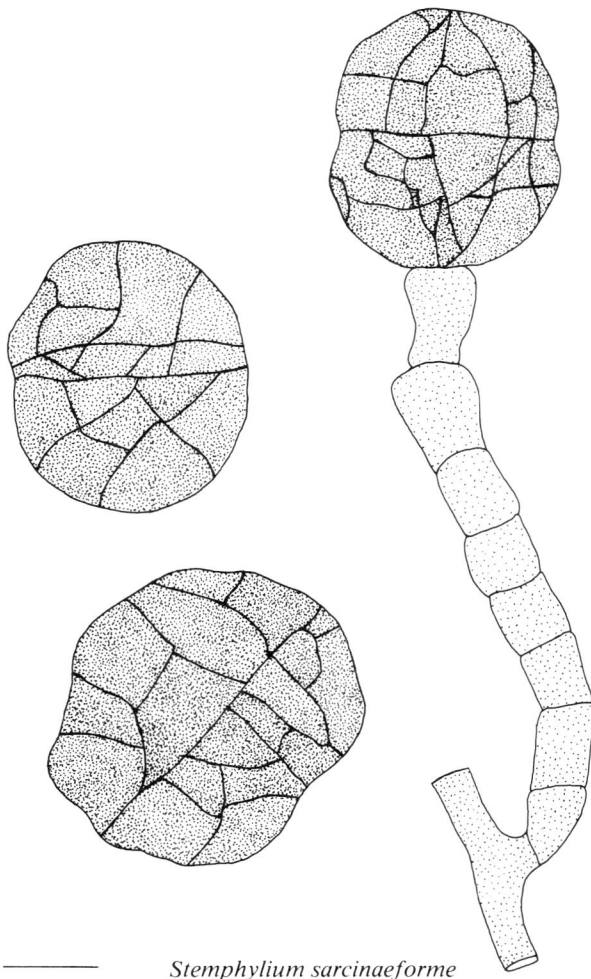

Stemphylium sarcinaeforme

Stemphylium Wallroth, 1833

Description: Colonies are moderately rapid growing, spreading, cottony, and light brown or olive green to black in color. Conidiophores are dematiaceous, simple or branched, percurrent, septate, with a swollen apex upon which the conidia form. Conidia are muriform, solitary, light brown to black, typically with a central constriction, rough or smooth walled.

Salient characteristics: *Stemphylium* is characterized by forming muriform conidia and percurrent proliferating conidiophores. *Stemphylium* is differentiated from *Ulocladium* by producing percurrent conidiophores.

Laboratory precautions: Handle with care, but special precautions are not necessary.

Key reference:
Simmons, 1967

107

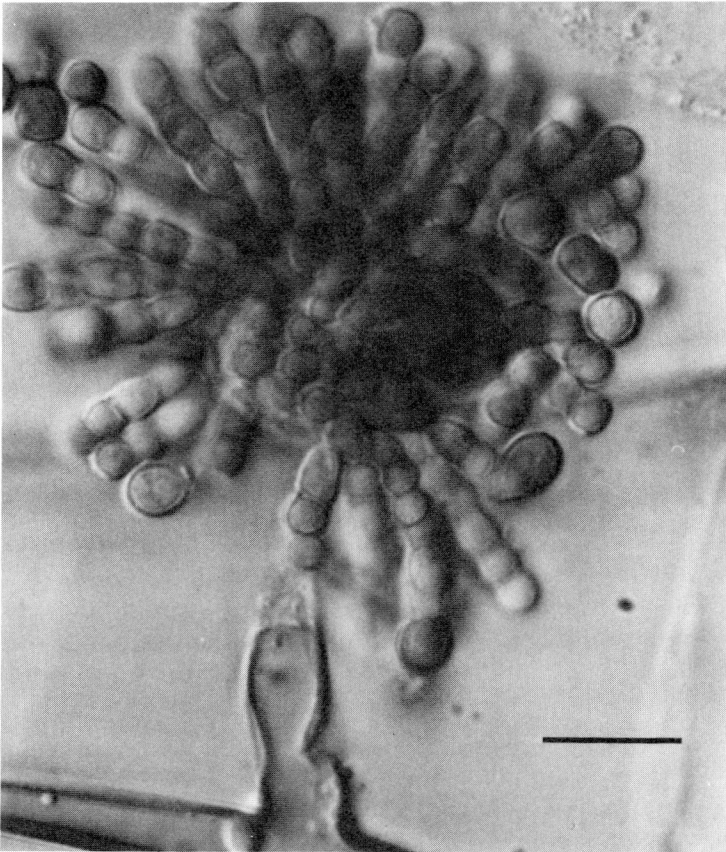

Syncephalastrum Schröter, 1886

Description: Colonies are rapid growing, often filling the culture tube, cottony, at first white, becoming gray to black, reverse color usually is white. Sporangiophores are branched, curved, and end in a vesicle. Merosporangia arise from the vesicles. Sporangiospores are 1-celled and globose. Rhizoids are usually present.

Salient characteristics: *Syncephalastrum* produces merosporangia that arise from vesicles at the apices of its sporangiophores.

Laboratory precautions: Handle with care, but special precautions are not necessary.

Key reference:
Zycha, Siepmann, and Linnemann, 1969

Figure 77. *Syncephalastrum racemosum.* The merosporangia are radiating from a vesicle at the apex of the sporangiophore. Bar is 10 μm.

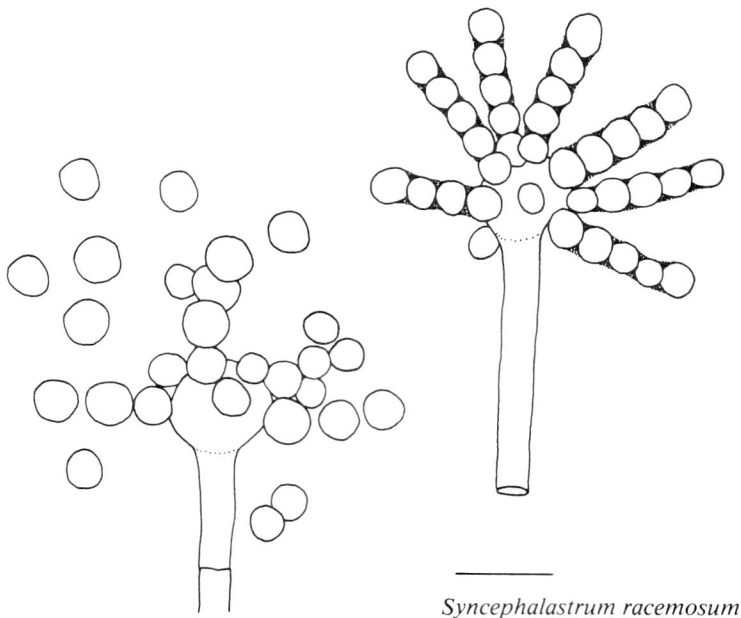

Syncephalastrum racemosum

108

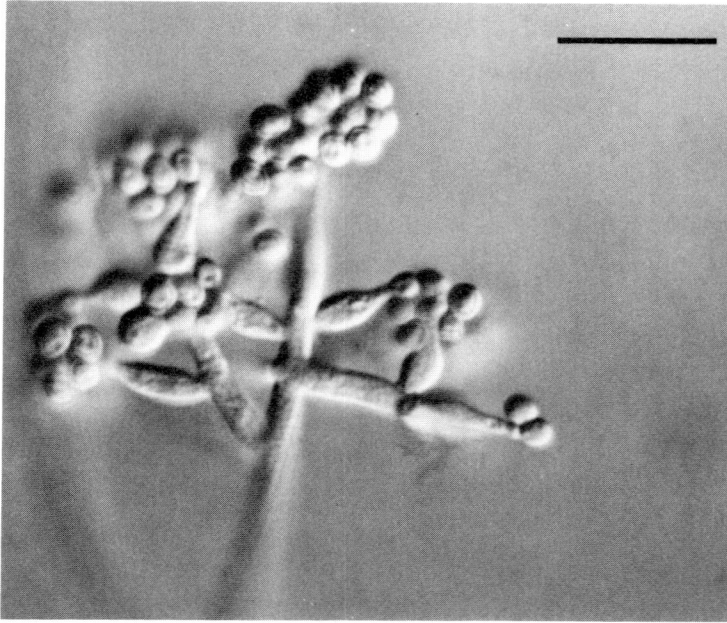

Figure 78. *Trichoderma viride.* The phialides arise at wide angles to the conidiophores. Bar is 10 μm.

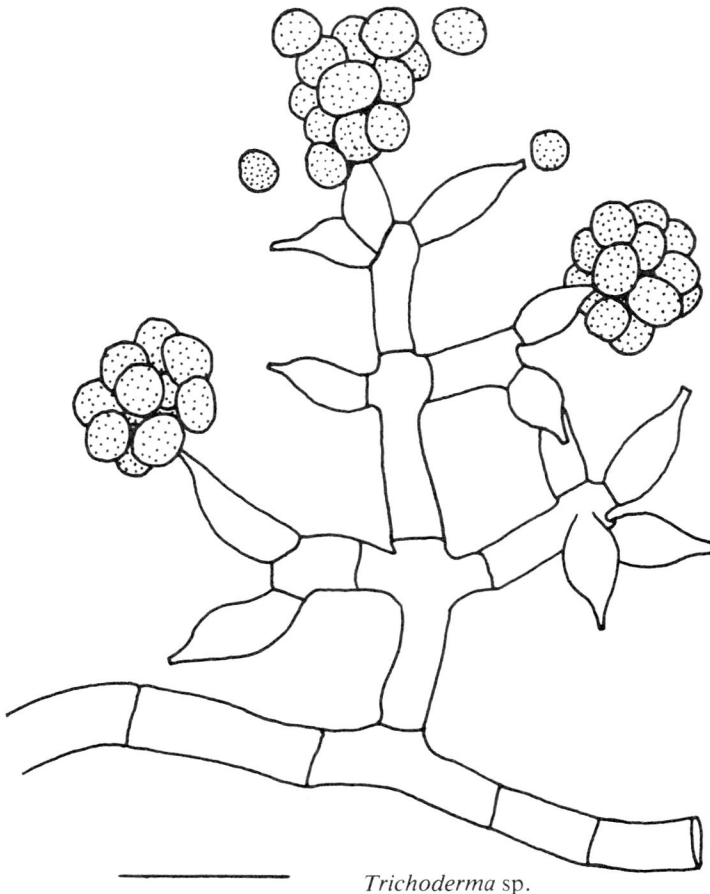

Trichoderma sp.

Trichoderma Persoon ex Gray, 1821

Description: Colonies are rapid growing, at first white and flat, becoming cottony and compact with dark green tufts; conidial areas occur as green, ringlike zones. Conidiophores are solitary or form compact tufts, are usually erect, arise at wide angles to the vegetative hyphae, and form concentric ringlike zones. Phialides are flask shaped, swollen in the central portion, taper toward the apex, are solitary or in clusters, hyaline, and arise at wide angles to the conidiophores. Conidia are 1-celled, subglobose to oblong, smooth or echinulate, hyaline to green (more common), and occur in balls at the apices of the phialides.

Salient characteristics: Isolates of *Trichoderma* form cottony to tufted, rapid-growing colonies that have green concentric ringlike zones, flask-shaped phialides that taper toward their apices and arise in irregular clusters at wide angles to their conidiophores, and 1-celled, usually green colored conidia that accumulate in balls.

Laboratory precautions: Handle with care, but special precautions are not necessary.

Key reference:
Rifai, 1969

109

Trichophyton Malmsten, 1845

Description: Colonies are slow to rapid growing, waxy, glabrous to cottony, white to brightly colored. Conidiophores are hyphalike. Macroconidia are 2- or more celled, generally thin walled, in some species thick walled, occasionally rare or absent, smooth, solitary, cylindrical, or clavate to cigar shaped. Microconidia are 1-celled, smooth, thin walled, hyaline, ovoid to clavate, solitary or in clusters. The microconidia are often the predominant type of conidia.

Salient characteristics: Macroconidia that are cylindrical, clavate to cigar shaped, thin or thick walled, smooth, and ovoid to clavate are characteristic of the genus *Trichophyton*. *Trichophyton* differs from *Microsporum* and *Epidermophyton* by having cylindrical, clavate to cigar-shaped, thin-walled or thick-walled, smooth macroconidia.

Laboratory precautions: Handle with care, but special precautions are not necessary.

Key references:
Dvořák and Otcenásek, 1969
Rebell and Taplin, 1970

Key to the More Common Species of *Trichophyton*

1.   *In vitro* hair perforation test positive 2
1.′   *In vitro* hair perforation test negative........................ 4
    2.   Colonies orange buff to tan with a purple black reverse and diffusible pigment; macroconidia thick walled, long, cylindrical to fusiform, slender, abundant, 8- to 12-celled .... *T. ajelloi*
    2.′   Colonies white to cream; macroconidia, when present, neither thick walled, long, nor slender 3
3.   Growth occurs at 37°C; macroconidia usually rare, 2- to 5-celled, thin walled, club shaped to cigar shaped; microconidia typically numerous, 1-celled, globose, solitary, along hyphae, or in clusters........................ *T. mentagrophytes*
3.′   Growth absent at 37°C; macroconidia 2- to 6-celled, thin walled, cylindrical; microconidia elongate to pyriform, typically 1-celled, rarely 2- to 3-celled, forming a transition from 1- to many-celled conidia .... *T. terrestre*

4. Colonies granular to woolly, often with abundant mycelium in the medium, folded, bright yellow, tan, or purplish red with a yellowish to mahogany red reverse; macroconidia rare, clavate to cigar shaped, thin walled, up to 10-celled; microconidia always numerous, arising at right angles to hyphae, globose, swollen, elongate, occasionally on matchstick-like conidiophores; growth enhanced by thiamine .... *T. tonsurans*

4.′ Characteristics not as above 5

5. Colonies rapid growing....................................... 6

5′ Colonies slow growing 7

6. Colonies white to pink with dark blood red reverse; requires L-histidine for growth; does not require nicotinic acid .... *T. megninii*

6.′ Colonies cream white to yellow, with reddish to tan centers and deep yellow or reddish brown reverse; requires nicotinic acid for growth; does not require L-histidine.. *T. equinum*

7. Colonies leathery, deep purple red, with a deep purple to violet reverse; growth enhanced by thiamine .... *T. violaceum*

7.′ Colonies not deep purple red with a deep purple to violet reverse.. 8

8. Colonies waxy, highly convoluted and heaped, cracking the agar, off-white to cream; favic chandeliers often common; conidia absent............ *T. schoenleinii*

8.′ Colonies not waxy, highly convoluted, or heaped 9

9. Colonies flat, woolly, or granular to cottony, white to cream with a carmine, blood red or olivaceous reverse; microconidia clavate to pyriform .... *T. rubrum*

9.′ Colonies glabrous to velvety, wrinkled, raised, furrowed, or folded...10

10. Reflexing hyphae present; colonies have raylike edges, leathery, wrinkled, furrowed, raised, yellow to dark apricot with a dark yellow to brown reverse ... *T. soudanense*

10.′ Reflexing hyphae absent, colonies without raylike edges 11

11. Colonies glabrous to velvety, often membranous, cream white, becoming deep tan or chocolate brown; may have a dark diffusible pigment; macroconidia unknown .... *T. yaoundei*

11.′ Colonies glabrous, folded, heaped, wrinkled, white, yellow to salmon with an unpigmented or salmon reverse; macroconidia 4- to 7-celled with an elongate end, formed on medium containing thiamine; thiamine and inositol (not all isolates require inositol) required for growth; growth enhanced at 37°C .... *T. verrucosum*

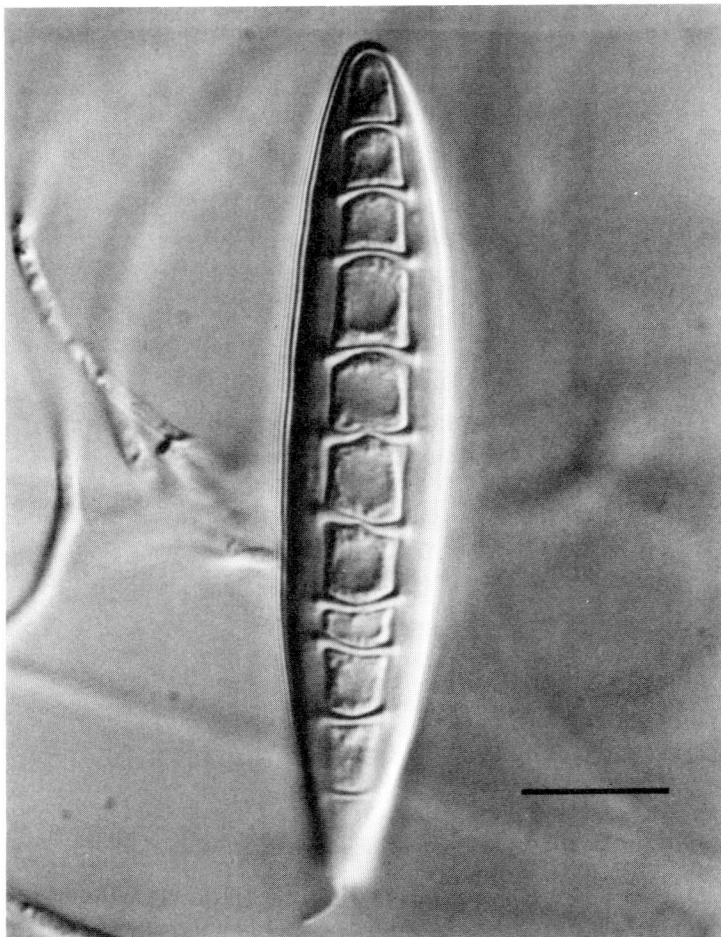

Trichophyton ajelloi
(Vanbreuseghem) Ajello, 1968

Description: Colonies are rapid growing, flat, sometimes with radial, shallow furrows, powdery to velvety, cream, yellow, buff, or with an apricot or yellow (sometimes dark purple) reverse color and a dark purple diffusible pigment. Macroconidia have 4 to 24 cells and are thick walled, smooth, long, and cylindrical to fusiform. Microconidia are sparse, 1-celled, ovoid to pyriform. *In vitro* hair perforation test is positive.

Salient characteristics: *Trichophyton ajelloi* produces a diffusible purple-black pigment from a velvety to powdery, yellow to orange colony having long, thick-walled, smooth, cigar-shaped macroconidia.

Figure 79. *Trichophyton ajelloi.* The macroconidium is smooth and has a thick wall. Bar is 10 μm.

Trichophyton equinum (Matruchot et Dassonville) Gedoelst, 1902

Description: Colonies are rapid growing, flat, velvety with some strains developing gentle folds, cream white to yellow with the centers frequently reddish, or orange brown to tan, with a deep yellow to deep reddish brown reverse color. Macroconidia rare, when present they are 2- to 6-celled, cylindrical to cigar shaped, thin walled. Requires nicotinic acid. Microconidia are 1-celled and ovoid to pyriform. *In vitro* hair perforation test is negative.

Salient characteristics: *Trichophyton equinum* is identified by the fact that it requires nicotinic acid, forms flat, velvety, white to orange brown colonies having a yellow to deep reddish brown reverse; cylindrical to cigar-shaped, 2-to 6-celled macroconidia; and forms 1-celled ovoid to pyriform microconidia. *Trichophyton equinum* differs from *T. mentagrophytes* by requiring nicotinic acid for growth and by not perforating hair.

Trichophyton megninii Blanchard, 1896

Description: Colonies are flat, with some gently, radially, irregularly furrowed or folded, velvety, and white to pink or pale purplish, with a dark blood red, rusty, or sienna reverse. Macroconidia are 1- to 8-celled, clavate to cigar shaped, thin walled. Microconidia are 1-celled and pyriform to clavate. Requires L-histidine. *In vitro* hair perforation test is negative.

Salient characteristics: *Trichophyton megninii* colonies require L-histidine and are flat, widely furrowed, white to pink with a claret red reverse. The species has rare cigar-shaped macroconidia with 2–8 cells, and 1-celled microconidia that are pyriform to clavate. *Trichophyton megninii* differs from *T. mentagrophytes* and *T. rubrum* by requiring L-histidine.

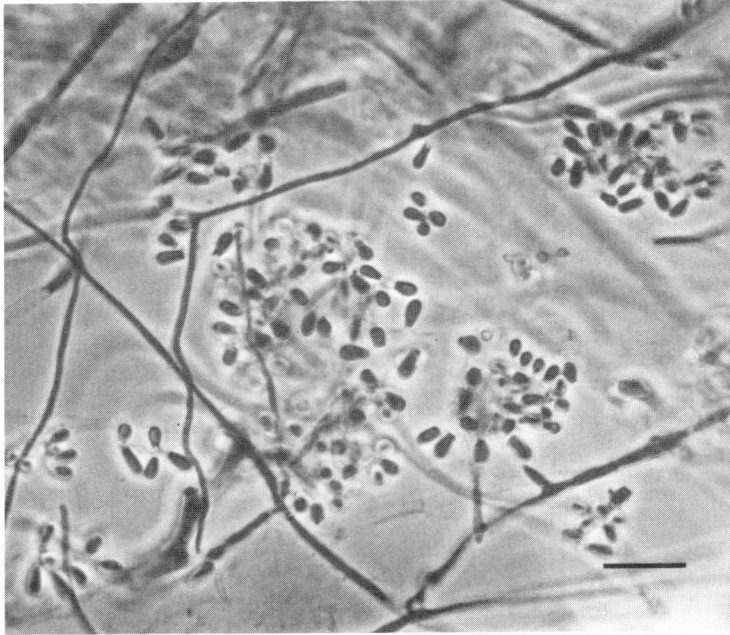

Figure 80. *Trichophyton mentagrophytes.* The microconidia are forming along the hyphae. Bar is 10 μm.

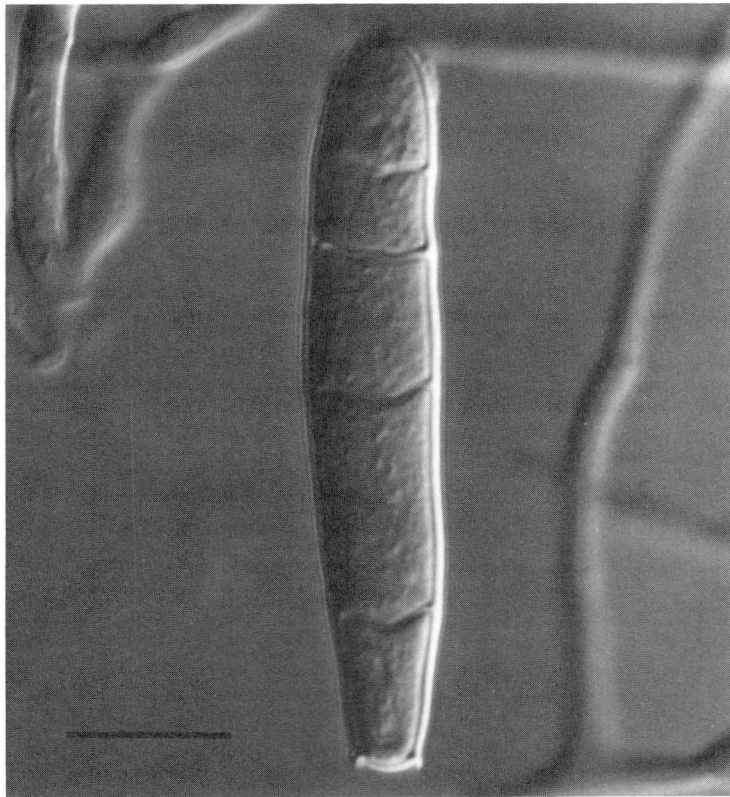

Figure 81. *Trichophyton mentagrophytes.* The macroconidium is cigar shaped. Bar is 10 μm.

Trichophyton mentagrophytes (Robin) Blanchard, 1896

Description: Colonies are moderately rapid growing, flat, powdery to downy, white, cream yellow, or cream buff. The powdery surface may have concentric rings. Colony reverse varies from yellow tan to cream, buff, red brown, brown, or deep wine red in color. Macroconidia are 3- to 8-celled, smooth, thin walled, clavate to cigar shaped, usually sparse. Microconidia are 1-celled, globose, arranged in grapelike clusters or along the hyphae. *In vitro* hair perforation test is positive.

Salient characteristics: *Trichophyton mentagrophytes* forms colonies that are flat, powdery to cottony, cream to buff or tan, with a reverse that is yellow, yellow brown, red brown, or wine red. The species also forms thin-walled, pencil- to cigar-shaped macroconidia, and globose, 1-celled microconidia. *Trichophyton mentagrophytes* differs from *T. rubrum* by perforating hair; from *T. equinum* by not requiring nicotinic acid for growth; and from *T. terrestre* by being able to grow at 37°C.

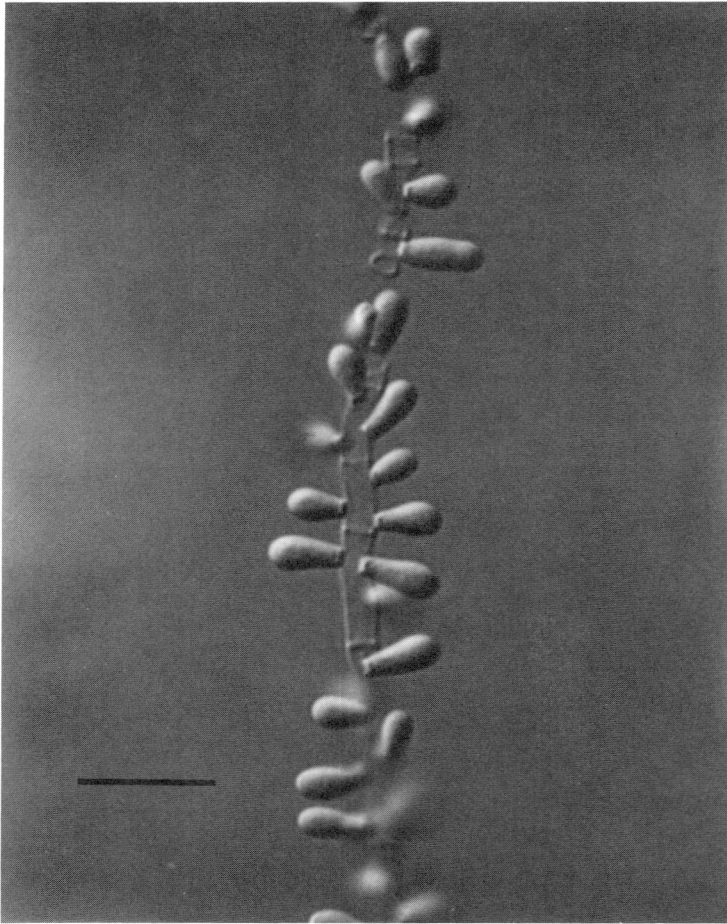

Figure 82. *Trichophyton rubrum.* The microconidia are clavate. Bar is 10 μm.

Trichophyton rubrum (Castellani) Sabouraud, 1911

Description: Colonies are slow growing and granular to cottony. Some isolates are suedelike to glabrous and heaped, white, cream, sometimes pale rose to mallow purple, with a carmine, dark blood red to olive reverse. The reverse color is enhanced by growth on vegetable polysaccharide agars containing dextrose. Macroconidia are 3- to 11-celled, sparse, long, narrow, cylindrical to cigar shaped, but usually poorly differentiated from hyphae. Microconidia are 1-celled, ovoid to clavate to pyriform, occurring only along the hyphae. *In vitro* hair perforation test is negative.

Salient characteristics: *Trichophyton rubrum* forms slow-growing, granular to cottony, heaped to domelike colonies with a dark red reverse; 3- to 11-celled, poorly differentiated macroconidia; and 1-celled, clavate to pyriform microconidia that develop along the hyphae. *Trichophyton rubrum* differs from *T. mentagrophytes* by being unable to perforate hair; it differs from *T. megninii* by not requiring L-histidine. *Trichophyton rubrum* usually forms a bright red pigment when grown on cornmeal agar containing 1% dextrose.

Trichophyton schoenleinii (Lebert) Langeron et Milochevitch, 1930

Description: Colonies are slow growing, waxy, slightly powdery or velvety, gently folded at first, becoming highly convoluted and heaped with age (often cracking and splitting the agar), off-white to cream. Reverse is without color or pale yellow. A diffusible pigment may be formed. Macroconidia and microconidia are absent (rarely present). Antlerlike hyphae with swollen clublike tips are often present. *In vitro* hair perforation test is negative.

Salient characteristics: *Trichophyton schoenleinii* forms sterile, waxy colonies that are folded or highly distorted and heaped, and that crack and split the agar. Most isolates produce favic chandeliers in the agar. *Trichophyton schoenleinii* differs from *T. verrucosum* by growing equally well at 25 and 37°C, and by not requiring inositol or thiamine for growth.

Trichophyton soudanense Joyeux, 1912

Description: Colonies are slow growing, with fringed or raylike edges, wrinkled, furrowed and folded, with a raised, heaped center having a leathery texture. They are yellow to dark apricot in color, with a dark yellow to brown reverse. Macroconidia are absent. Microconidia are 1-celled, ovoid to pyriform, borne laterally on the hyphae, occasionally in clusters. Hyphae branch at right angles and reflex backward. *In vitro* hair perforation is negative.

Salient characteristics: Isolates of *T. soudanense* are slow growing, heaped, leathery, and yellow to apricot in color with a dark yellow reverse. They lack macroconidia, but may have pyriform microconidia and reflexing hyphae.

116

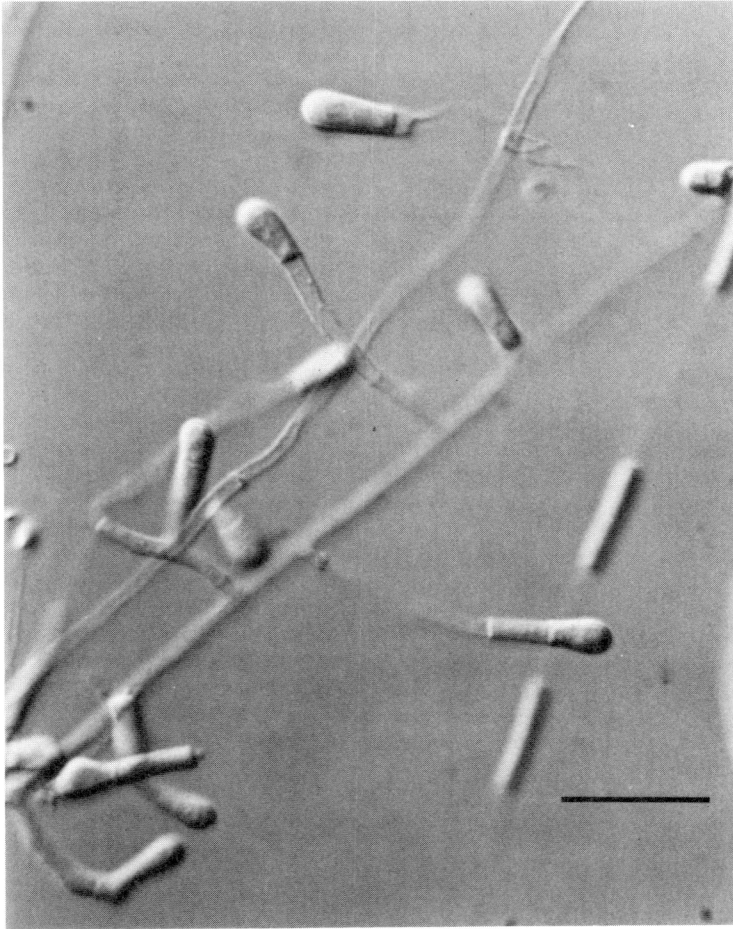

Figure 83. *Trichophyton terrestre.* The microconidia have a broad base. Bar is 10 μm.

Trichophyton terrestre
Durie et Frey, 1957

Description: Colonies are rapid growing, velvety to granular, white, buff, yellow to pink red, with a yellow to yellow greenish reverse color. Macroconidia are 4- to 6-celled, smooth, thin, cylindrical. Microconidia are typically long, cylindrical, ovoid to cigar shaped. All conidia show a prominent attachment scar and stain more intensely with lactophenol cotton blue than do the hyphae. In red forms of *T. terrestre,* the microconidia are numerous; the macroconidia are absent or rare. *In vitro* hair perforation test is positive.

Salient characteristics: *Trichophyton terrestre* forms colonies that are velvety to granular and white to yellow, with a red to yellow reverse. The species also forms abundant ovoid to cigar-shaped macroconidia having 2 to 4 cells. *Trichophyton terrestre* differs from *T. mentagrophytes* by not being able to grow at 37°C; it differs from *T. rubrum* by perforating hair; and from *T. tonsurans* by not having its growth enhanced by thiamine.

117

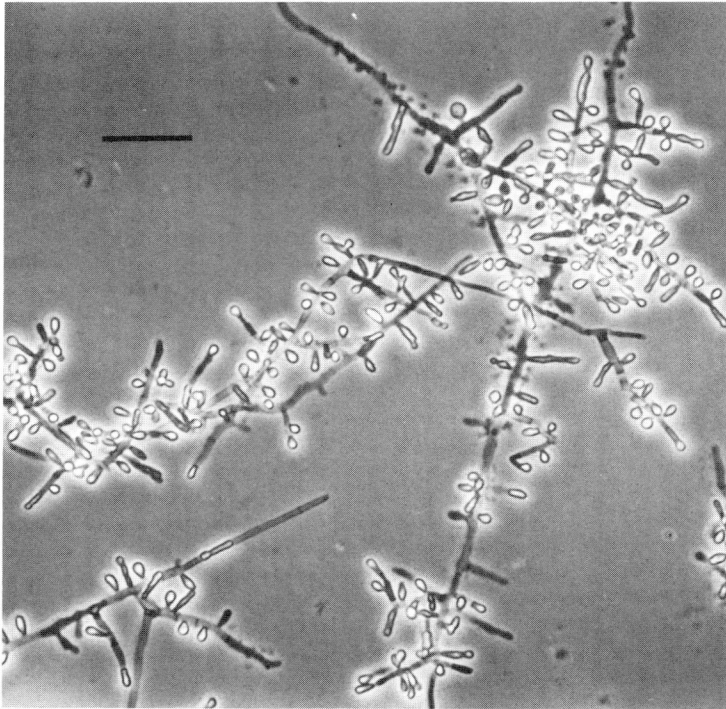

Figure 84. *Trichophyton tonsurans.* The conidia are extremely variable. Bar is 10 μm.

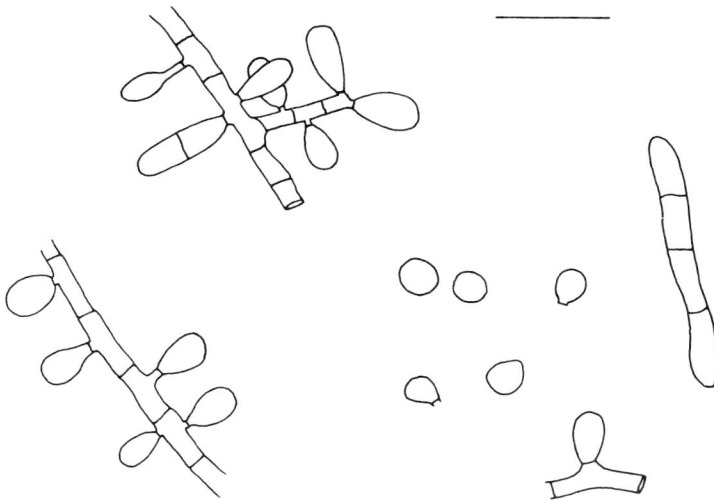

Trichophyton tonsurans

Trichophyton tonsurans
Malmsten, 1845

Description: Colonies are slow growing, of four colony types: crateriform, cerebriform, plicatile, or flat. Predominant colony type is flat, granular, becoming velvety, wrinkled, white to yellowish with a yellow brown, rust, or mahogany reverse color. Macroconidia are 2- to 10-celled, clavate to cylindrical, smooth, thin walled, with their apices slightly bent to one side. Microconidia are 1-celled, abundant, variable in size and shape, clavate, teardrop to subglobose, or swollen like balloons, borne on elongate conidiophores that arise at right angles to the parent hypha. The branchings of hyphae that bear the conidia are often at right angles and sometimes thickened. Growth is enhanced by thiamine. *In vitro* hair perforation test is negative.

Salient characteristics: Colonies of *T. tonsurans* are granular, flat, furrowed, yellow to red brown, and have abundant microconidia ranging from clavate to teardrop to balloonlike in shape, with some conidia borne on matchsticklike conidiophores. Growth is enhanced by thiamine.

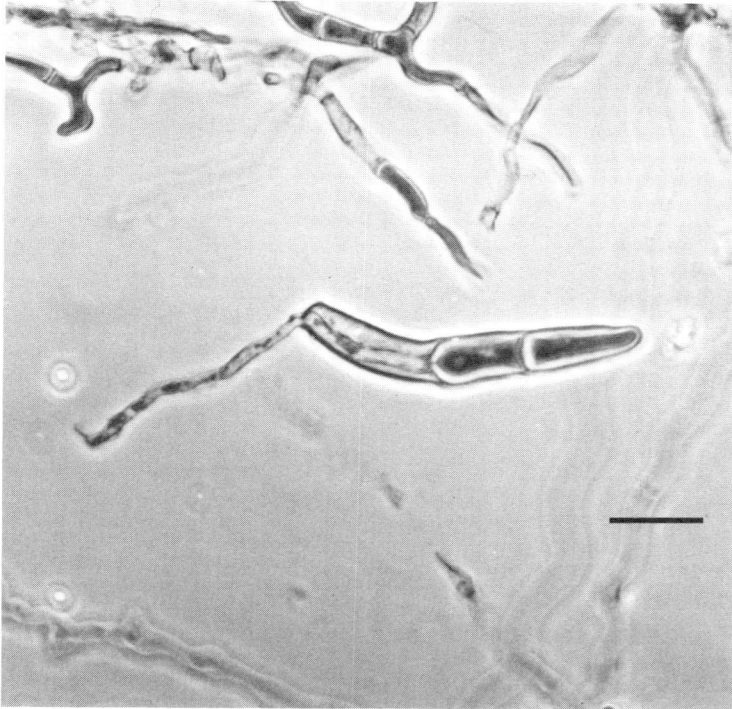

Figure 85. *Trichophyton verrucosum.* A macroconidium. Bar is 10 μm.

Trichophyton verrucosum
Bodin, 1902

Description: Colonies are slow growing, at first glabrous, becoming knoblike, slightly folded, wrinkled, raised, white, sometimes salmon to yellow. Reverse is unpigmented or salmon. Macroconidia are rare, 4- to 7-celled, smooth, thin walled, fusiform, with one end tapering like the tail of a rat (formed on enriched media with thiamine and inositol). Microconidia are 1-celled, ovoid to pyriform, but typically absent. Favic chandeliers often are present. Growth is enhanced at 37°C. The species require thiamine (some isolates require inositol also) for growth. *In vitro* hair perforation test is negative.

Salient characteristics: *Trichophyton verrucosum* forms colonies that are white, slow growing, and buttonlike with rare macroconidia and pyriform microconidia. Isolates grow better at 37°C than at 30°C and require thiamine and inositol (some isolates). *Trichophyton verrucosum* differs from *T. schoenleinii* by having enhanced growth at 37°C and by requiring inositol (some isolates) and thiamine for growth.

Trichophyton violaceum Sabouraud in Bodin, 1902

Description: Colonies are extremely slow growing, leathery, wrinkled, heaped, verrucous, glabrous, deep purple red (somewhat paler in downy variants), with a deep purple to violet reverse color. Macroconidia are rare, 2-to 8- celled, smooth and irregular in shape. Microconidia are 1-celled, rare, and ovoid to pyriform. Development of conidia is enhanced on thiamine-enriched media. Hyphae are highly distorted. Growth is enhanced by thiamine. *In vitro* hair perforation test is negative.

Salient characteristics: Isolates of *T. violaceum* are extremely slow growing, wrinkled, red to purplish, with a reverse color that is red to purple violet. Conidia are absent. *Trichophyton violaceum* differs from other faviform *Trichophyton* species by having its growth enhanced by thiamine and by producing a purple violet pigment.

Trichophyton yaoundei Cochet et Doby-Dubois, 1957

Description: Colonies are slow growing, glabrous to velvety, often membranous, raised, with a tendency to form folds, cream white, becoming deep tan to chocolate brown with age (a dark pigment may diffuse into the medium), reverse is unpigmented. Macroconidia are absent. Microconidia are rare, 1-celled, and ovoid to pyriform. Favic chandeliers may be numerous. *In vitro* hair perforation test is negative.

Salient characteristics: *Trichophyton yaoundei* colonies are glabrous, raised, folded, cream white, turning dark brown with age, and forming a pigment that diffuses into the medium; rare pyriform microconidia and favic chandeliers may be formed.

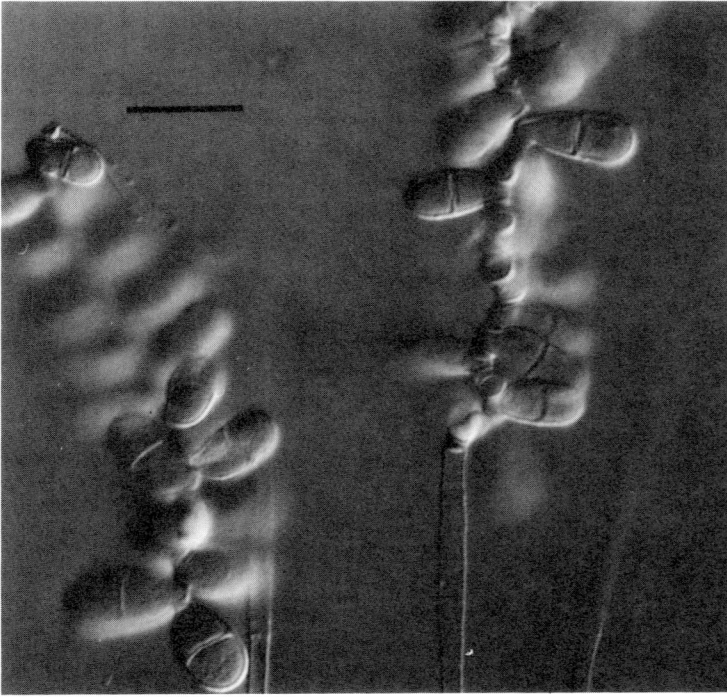

Figure 86. *Trichothecium roseum.* The 2-celled clavate conidia alternate with each other to form a chain. Bar is 10 μm.

Trichothecium Link ex Gray, 1821

Description: Colonies are rapid growing, granular, and pale rose in color. Conidiophores are hyphalike, erect, solitary or in clusters, branched or simple, hyaline, septate, becoming shorter as the conidia develop. Conidia are 2-celled, clavate to ovoid, hyaline to lightly colored, alternating with each other to form chains.

Salient characteristics: Isolates of *Trichothecium* are pale rose, granular, and rapidly developing, with 2-celled, clavate conidia that alternate with each other to form chains at the apices of the conidiophores.

Laboratory precautions: Handle with care, but special precautions are not necessary.

Key reference:
Rifai and Cooke, 1966

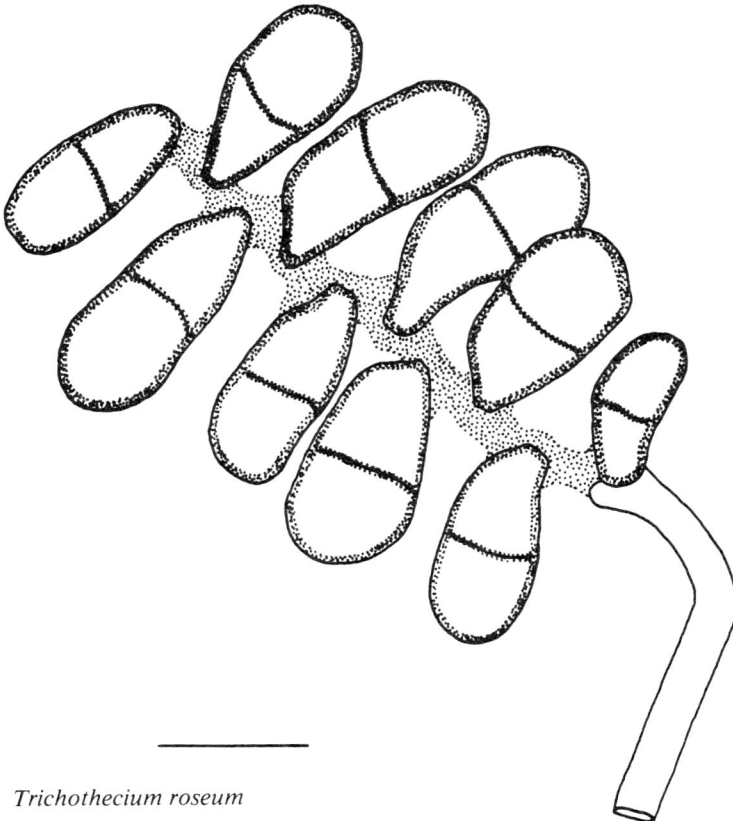

Trichothecium roseum

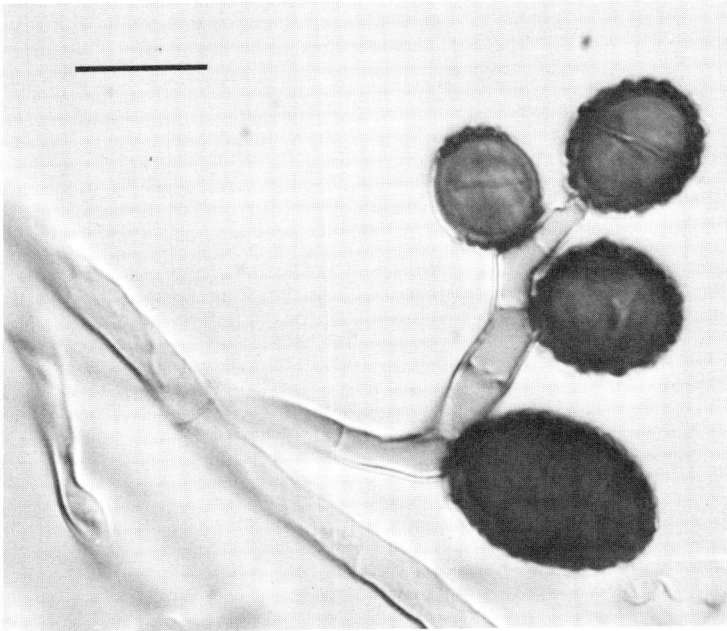

Figure 87. *Ulocladium atrum* . The muriform conidia are developing from a geniculate conidiophore. Bar is 10 μm.

Ulocladium Preuss, 1851

Description: Colonies are moderately rapid growing, velvety, brown or olivaceous brown to black, partly immersed in the medium. Conidiophores are simple or branched, brown, septate, sympodial, and geniculate in appearance. Conidia are muriform, obovoid, solitary, smooth or rough, and brown to black in color.

Salient characteristics: Isolates of *Ulocladium* form geniculate, sympodial conidiophores bearing solitary, muriform, dematiaceous conidia. *Ulocladium* is differentiated from *Stemphylium* by having geniculate, sympodial conidiophores.

Laboratory precautions: Handle with care, but special precautions are not necessary.

Key reference:
Simmons, 1967

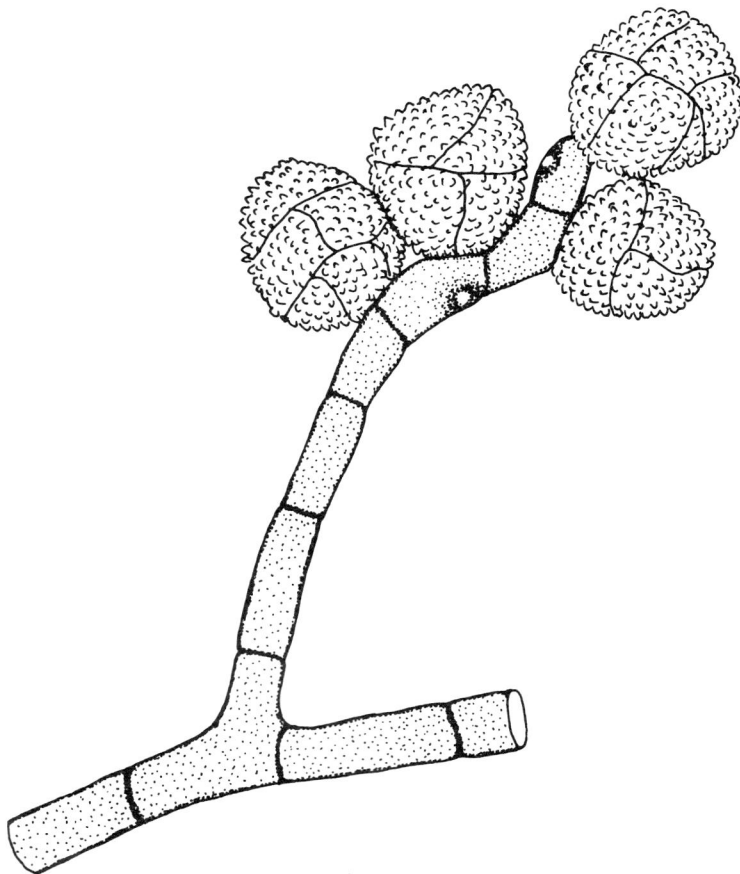

Ulocladium atrum

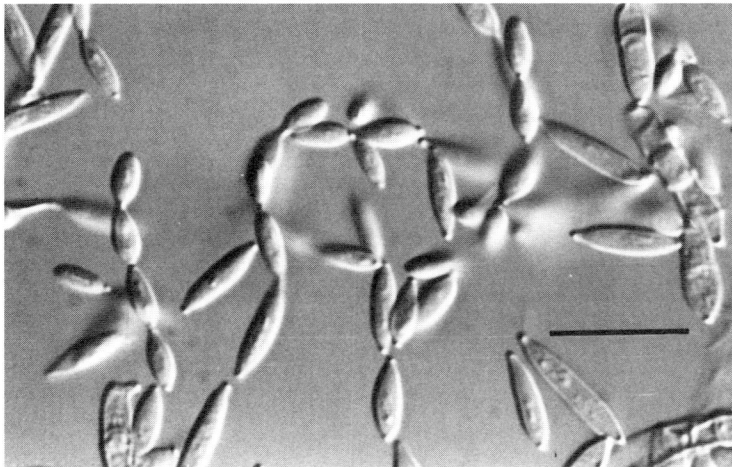

Figure 88. *Ustilago violaceum.* The blastoconidia are spindle shaped. Bar is 10 μm.

Ustilago violaceum

Ustilago (Persoon) Roussel, 1806

Description: Colonies are slow growing, moist, yeastlike, becoming wrinkled and membranelike, tan to dark brown in color, with profuse budding in the medium. Hyphae are present. Yeast form consists of elongate, spindle-shaped cells that give rise to similar blastoconidia. Pseudohyphae consisting of spindle-shaped cells are often formed. Teliospores may be present.

Salient characteristics: Tan to brown, wrinkled, yeastlike colonies having spindle-shaped parent cells with similar blastoconidia are suggestive of *Ustilago* and of several similar species of smuts. Members of this genus cannot be identified in the clinical laboratory without additional phytopathologic studies.

Laboratory precautions: Handle with care, but special precautions are not necessary.

Key reference:
Ainsworth and Sampson, 1950

123

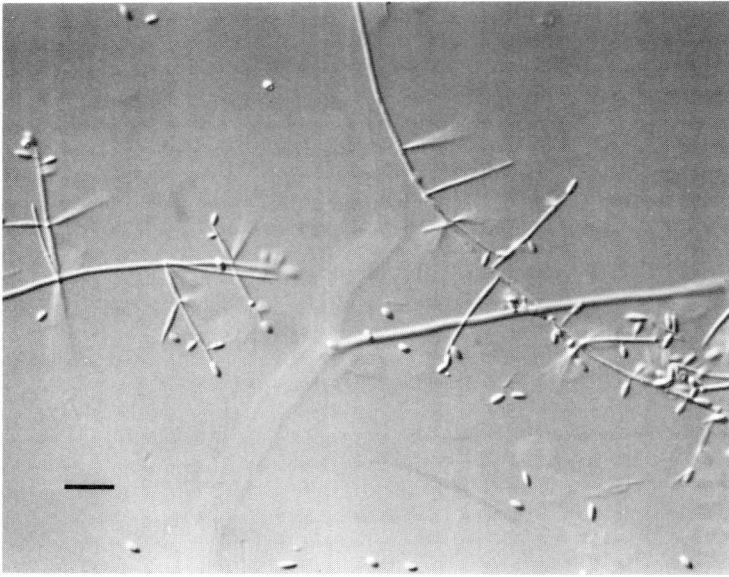

Figure 89. *Verticillium* sp. The phialides arise in whorls, upon which balls of conidia accumulate at their apices. Bar is 10 μm.

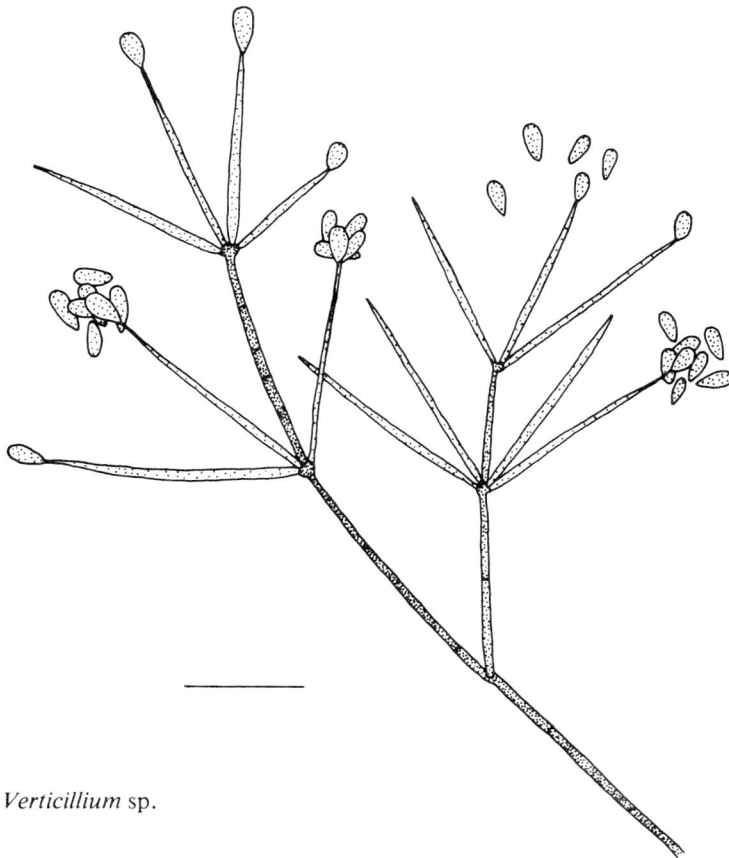

Verticillium sp.

Verticillium Nees ex Steudel, 1824

Description: Colonies are rapid growing, spreading, velvety to cottony, white, green, yellow, red, or pinkish brown in color. Microsclerotia are often present. Conidiophores are erect, hyaline to pigmented, simple, or branched. The branching occurs in whorls at several levels. Phialides are lageniform, hyaline, and arise in whorls at the apices of the branches or along the conidiophores. Conidia are 1-celled, hyaline, ovoid to allantoid, occurring as balls at the apices of the phialides.

Salient characteristics: *Verticillium* forms whorls of branches and lageniform phialides having balls of 1-celled conidia at their apices. *Verticillium* differs from *Acremonium* by having whorls of phialides and branches.

Laboratory precautions: Handle with care, but special precautions are not necessary.

Key reference:
Gams, 1971

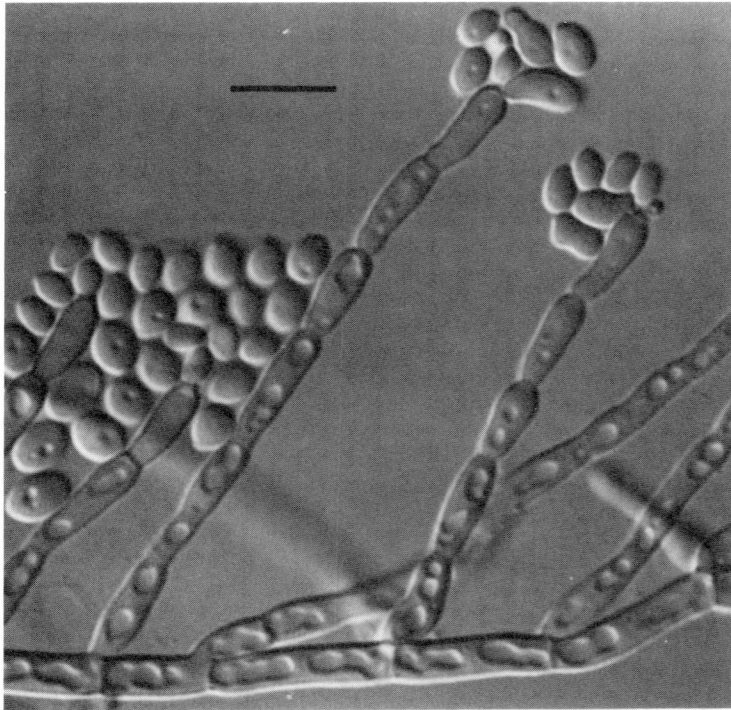

Figure 90. *Wangiella dermatitidis*. The phialides do not have collarettes. Bar is 10 µm. (Reproduced by permission from PAHO Scientific Publ. No. 356, pp. 37–59, 1978.)

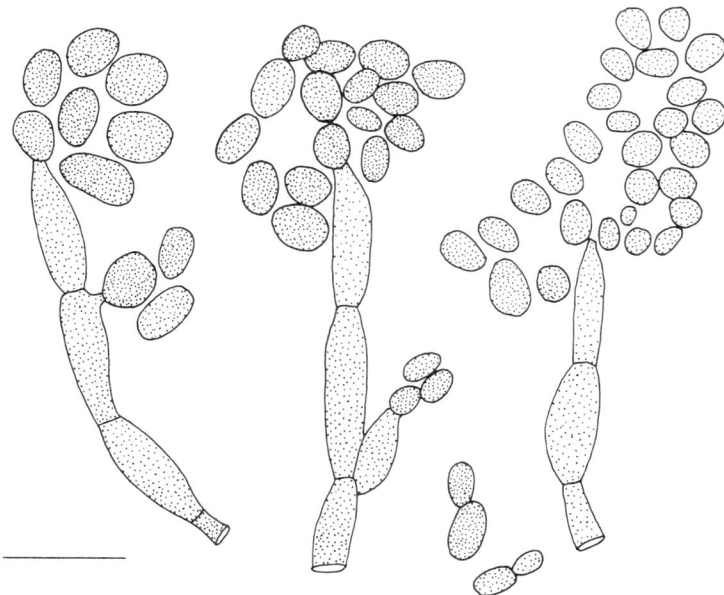

Wangiella dermatitidis

Wangiella McGinnis, 1977

Description: Colonies are slow growing, at first yeastlike and black, becoming velvety and olivaceous gray with an iron gray reverse color. Conidiophores are usually indistinguishable from the vegetative hyphae. Phialides are without collarettes, flask shaped, intercalary or terminal, light brown. Conidia are 1-celled, subglobose to obovoid, smooth, pale brown, and accumulating as balls at the apices of the phialides. A budding yeast form is common. Annellides also may be present. Growth occurs at 40°C.

Salient characteristics: Isolates of *Wangiella* produce yeastlike colonies that are sticky to the touch, becoming velvety, phialides without collarettes, and a budding yeast form. *Wangiella* differs from *Phialophora* by the absence of collarettes; it differs from *Exophiala* by producing phialides. It differs specifically from *E. jeanselmei* by growing at 40°C, as well as by having phialides without collarettes, compact conidiogenous cells, and a yeast form that does not have short, lateral (subapical portion) tubular denticles bearing annellations.

Laboratory precautions: Handle with care, but special precautions are not necessary.

Key reference:
McGinnis, 1980

SECTION B   Identification of Yeasts

The incidence of infections caused by species of yeasts has increased dramatically in recent years. This increase can be accounted for by medical advances that have extended the lives of severely ill patients, but have concomitantly increased their susceptibility to infections caused by endogenous and exogenous fungi. As a result, new therapeutic modalities have been developed to treat specific yeast infections; these modalities now require laboratories to identify yeasts to the species level.

In addition to the methods for yeast identification described in this handbook, numerous identification systems are available commercially. When they are used in conjunction with morphologic characteristics (and, when necessary, in conjunction with additional physiologic tests), these identification systems permit the rapid and accurate identification of the majority of yeasts encountered in the clinical laboratory. As with all microorganisms, the approach adopted by each laboratory for yeast identification should be as accurate, rapid, and economical as possible. Because *Candida albicans* is the yeast most frequently encountered in the clinical laboratory, a germ tube test should be performed initially on all clinically significant isolates. Isolates that are germ tube positive can be identified definitively as either *C. albicans* or *C. stellatoidea;* a sucrose assimilation test is necessary to distinguish between these two species. The identification of germ-tube-negative yeast isolates should be based upon both morphologic and physiologic characteristics (see Tables 2.2 and 2.3). The exclusion of either of these types of information can compromise severely the accuracy of the final identification. For these data to be meaningful, pure cultures must be studied.

Some authors have included members of the genus *Torulopsis* in the genus *Candida*. For reasons that are beyond the scope of this handbook, we have elected not to combine them. Additionally, we have accepted *Trichosporon penicillatum* as a member of the genus *Geotrichum*.

TABLE 2.2 Morphologic Characteristics of Selected Yeasts on Dalmau Plates

Organism	Morphology on Dalmau Plate[a]
Candida aaseri	Pseudohyphae short, rudimentary, and abundant; blastoconidia cylindrical.
C. albicans	Pseudohyphae well developed, abundant, often with clusters of globose to oval blastoconidia at the septa; chlamydospores usually produced; true hyphae may be present, especially in old isolates.
C. guilliermondii	Pseudohyphae rudimentary or well developed; when well developed the pseudohyphae are slender, often curved with ovoid or elongate blastoconidia in chains, chains somewhat verticilliate.
C. humicola	Pseudohyphae and true hyphae abundant, often wavy in appearance; branches of hyphae often parallel to main axes; blastoconidia not well differentiated.
C. krusei	Pseudohyphae consisting of long cells with treelike branching; verticils and chains of blastoconidia arising at points of branching; pseudohyphae curved with scarce blastoconidia in some isolates.
C. lambica	Pseudohyphae abundant, consisting of long, slender, branched cells; blastoconidia in short chains arising from the pseudohyphae.
C. lipolytica	Pseudohyphae and true hyphae abundant; single and paired blastoconidia form terminally around the pseudohyphae and true hyphae.
C. lusitaniae	Pseudohyphae abundant, slender and curved; blastoconidia forming short verticillated chains
C. parapsilosis	Pseudohyphae usually thin, branched, and usually bearing verticils of a few blastoconidia; pseudohyphae consisting of curved cells; giant cells may occur.
C. paratropicalis	Resembles *C. tropicalis.*
C. pseudotropicalis	Pseudohyphae typically abundant, branched; chains of blastoconidia often verticillate.
C. rugosa	Pseudohyphae abundant, primitive, and extremely branched; blastoconidia scarce.
C. stellatoidea	Resembles *C. albicans.*
C. tropicalis	Pseudohyphae abundant, long and branched; blastoconidia single or in short chains produced in verticils from the pseudohyphae; true hyphae may be present.
C. zeylanoides	Pseudohyphae abundant and robust; blastoconidia in lateral chains or verticillated groups.
Cryptococcus spp.[b]	Pseudohyphae absent (rudimentary if present); blastoconidia globose to oval, attached to parent cells by narrow necks, may possess capsules.
Geotrichum spp.[c]	True hyphae abundant; arthroconidia produced by fragmentation of hyphae; blastoconidia absent.

[a]Cornmeal — Tween 80 agar.
[b]Includes *Cryptococcus albidus, C. gastricus, C. laurentii, C. neoformans, C. terreus,* and *C. uniguttulatus.*
[c]Includes *Geotrichum candidum* and *G. penicillatum.* See mould section for generic description.

Organism	Morphology on Dalmau Plate[a]
Malassezia spp.[d]	Usually requires media containing lipid sources such as olive oil (*M. furfur* only); hyphae not usually present; small, flask-shaped cells produce phialoconidia; unipolar budding.
Rhodotorula spp.[e]	Pseudohyphae absent (rudimentary if present); blastoconidia globose to oval, may possess capsules.
Saccharomyces spp.[f]	Pseudohyphae may be present, usually absent or rudimentary; asci containing 1–4 ascospores present, but may require special media for their formation.
Sporobolomyces salmonicolor	True hyphae and pseudohyphae typically present; blastoconidia predominant; ballistoconidia sickle to kidney shaped, produced on denticles from the vegetative cells.
Torulopsis spp.[g]	Pseudohyphae absent; (rudimentary if present); blastoconidia and vegetative cells small.
Trichosporon spp.[h]	True hyphae abundant; pseudohyphae present in variable amounts; arthroconidia produced by fragmentation of hyphae; blastoconidia present, often few in number.

[d]Includes *Malassezia furfur* and *M. pachydermatis.*

[e]Includes *Rhodotorula glutinis, R. graminis, R. pilimanae,* and *R. rubra.*

[f]Includes *Saccharomyces cerevisiae* and *S. rouxii.*

[g]Includes *Torulopsis candida* and *T. glabrata.*

[h]Includes *Trichosporon beigelii, T. capitatum, T. fermentans,* and *T. pullulans.*

Source: Modified from Lodder (ed.). 1970. *The Yeasts: A Taxonomic Study.* North-Holland Publishing and McGinnis. 1980. *Laboratory Handbook of Medical Mycology.* Academic Press.

TABLE 2.3 Some Physiologic and Morphologic Characteristics of Selected Yeasts

Yeast	Cellobiose	Galactitol[b]	Galactose	Glucose	Inositol	Lactose	Maltose	Melezitose	Melibiose	Raffinose	Sorbose	Sucrose	Trehalose	Xylose	Nitrate Assimilation	Urea Hydrolysis	Germ Tube	Chlamydospores	Strong Phenoloxidase Activity
Candida aaseri	+[c]	−	+	+	−	+	+	+	−	−	+	+	+	+	−	−	−	−	−
C. albicans	−[d]	−	+	+	−	−	+	V	−	−	V	+	+	+	−	−	+	+	−
C. guilliermondii	+	+	+	+	−	−	+	+	+	+	+	+	+	+	−	−	−	−	−
C. humicola	+	V	+	+	+	+	+	V	+	V	V	+	+	+	−	+	−	−	−
C. krusei	−	−	−	+	−	−	−	−	−	−	−	−	−	−	−	V	−	−	−
C. lambica	−	−	−	+	−	−	−	−	−	−	−	−	+	−	−	−	−	−	−
C. lipolytica	−	−	−	+	−	−	−	−	−	−	V	−	−	−	−	V	−	−	−
C. lusitaniae	+	−	+	+	−	−	+	+	−	−	+	+	+	+	−	−	−	−	−
C. paratropicalis	V	−	+	+	−	−	+	V	−	−	+	V	+	+	−	−	−	−	−
C. parapsilosis	−	−	+	+	−	−	+	+	−	−	+	+	+	+	−	−	−	−	−
C. pseudotropicalis	+	−	+	+	−	+	−	−	−	+	−	+	−	+	−	−	−	−	−
C. rugosa	−	−	+	+	−	−	−	−	−	−	V	−	−	+	−	−	−	−	−
C. stellatoidea	−	−	+	+	−	−	+	−	−	−	−	+	+	−	−	−	+	+	−
C. tropicalis	V[e]	−	+	+	−	−	+	+	−	−	V	+	+	+	−	−	−	−	−
C. zeylanoides	−	−	V	+	−	−	−	−	−	−	+	−	+	−	−	−	−	−	−
Cryptococcus albidus	+	V	+	+	+	+	+	+	−	+	V	+	+	+	+	+	−	−	−
C. gastricus	+	−	+	+	+	V	+	+	−	−	−	+	+	+	−	+	−	−	−
C. laurentii	+	+	+	+	+	+	+	+	+	+	V	+	+	+	−	+	−	−	−
C. neoformans	+	+	+	+	−	+	+	+	−	+	V	+	+	+	−	+	−	−	+
C. terreus	+	V	+	+	+	+	V	V	−	−	+	−	+	+	+	+	−	−	−
C. uniguttulatus	V	−	V	+	+	−	+	+	−	V	−	+	V	+	−	+	−	−	−
Geotrichum spp.[a]	−	−	+	+	−	−	−	−	−	−	+	−	+	−	−	−	−	−	−
Rhodotorula glutinis	+	−	V	+	−	−	+	+	−	V	V	+	+	+	+	+	−	−	−
R. graminis	V	+	+	+	−	−	−	−	−	−	V	+	+	+	+	+	−	−	−
R. pilimanae	V	−	+	+	−	−	−	−	−	−	+	+	+	+	+	+	−	−	−
R. rubra	V	−	V	+	−	−	+	+	−	+	V	+	+	+	+	+	−	−	−
Saccharomyces cerevisiae	−	−	+	+	−	−	+	V	+	+	−	+	V	−	−	−	−	−	−
S. rouxii	−	−	+	+	−	−	+	−	−	−	−	V	V	−	−	−	−	−	−
Sporobolomyces salmonicolor	−	−	−	+	−	−	−	−	−	+	V	+	+	−	+	+	−	−	−
Torulopsis candida	+	V	+	+	−	V	+	V	+	−	+	+	+	+	−	−	−	−	−
T. glabrata	−	−	+	+	−	−	−	−	−	−	−	−	+	−	−	−	−	−	−
Trichosporon beigelii	+	V	+	+	+	+	+	V	V	V	V	+	V	+	−	+	−	−	−
T. capitatum	−	−	+	+	−	−	−	−	−	−	V	−	−	−	−	−	−	−	−
T. fermentans	+	−	+	+	−	−	−	−	−	+	−	−	+	−	−	+	−	−	−
T. pullulans	+	−	+	+	V	+	+	+	+	+	V	+	+	V	+	+	−	−	−

[a] Includes *Geotrichum candidum* and *G. penicillatum*. See mould section for generic description.

[b] also known as dulcitol.

[c] +: ≥ 90% of strains positive.

[d] −: ≥ 90% of strains negative.

[e] V: variable.

Source: Modified from Lodder (ed.). 1970. *The Yeasts: A Taxonomic Study.* North Holland Publishing and McGinnis. 1980. *Laboratory Handbook of Medical Mycology.* Academic Press.

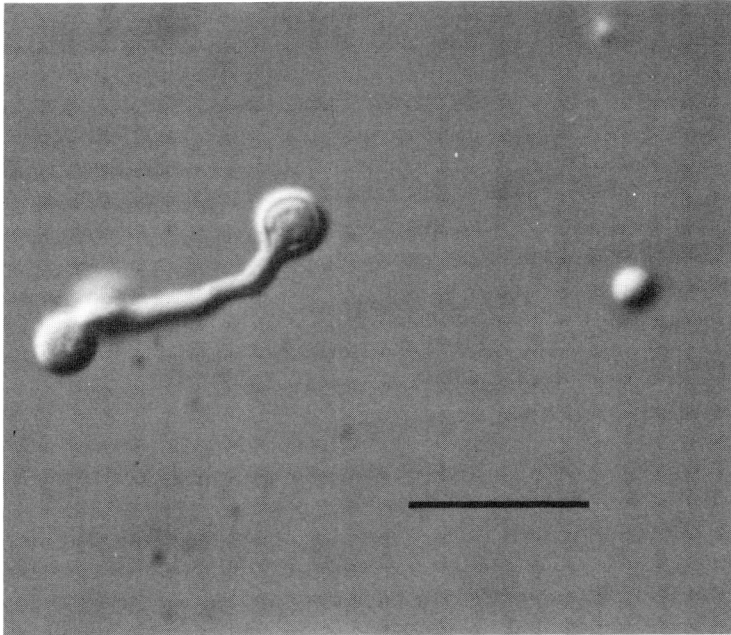

Figure 91. *Candida albicans.* The germ tube was produced in human serum after 3-hr. incubation at 37°C. Bar is 10 μm.

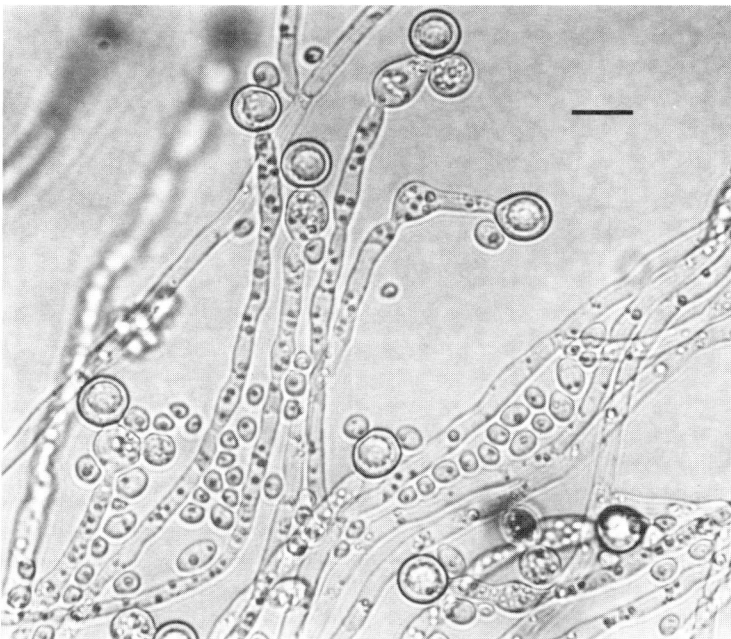

Figure 92. *Candida albicans.* The chlamydospores developed on yeast morphology agar inoculated by the Dalmau technique. Bar is 10 μm.

Candida Berkhout, 1923 *nom. cons.*

Description: Colonies are rapid growing, soft, glistening or dull, smooth or wrinkled, white to yellowish cream in color. Pseudohyphae are well developed. Hyphae may be formed. Blastoconidia are 1-celled, globose to cylindrical, sometimes irregular in shape, with multipolar budding. Gaseous fermentation occurs in many species. Sexual spores and arthroconidia are absent.

Salient characteristics: Well developed pseudohyphae and 1-celled blastoconidia characterize the common species of *Candida. Candida* differs from *Cryptococcus* and *Torulopsis* by having well-developed pseudohyphae; it differs from *Trichosporon* by not producing arthroconidia.

Laboratory precautions: Handle with care, but special precautions are not necessary.

Key references:
Barnett, Payne, and Yarrow, 1979
Lodder, 1970

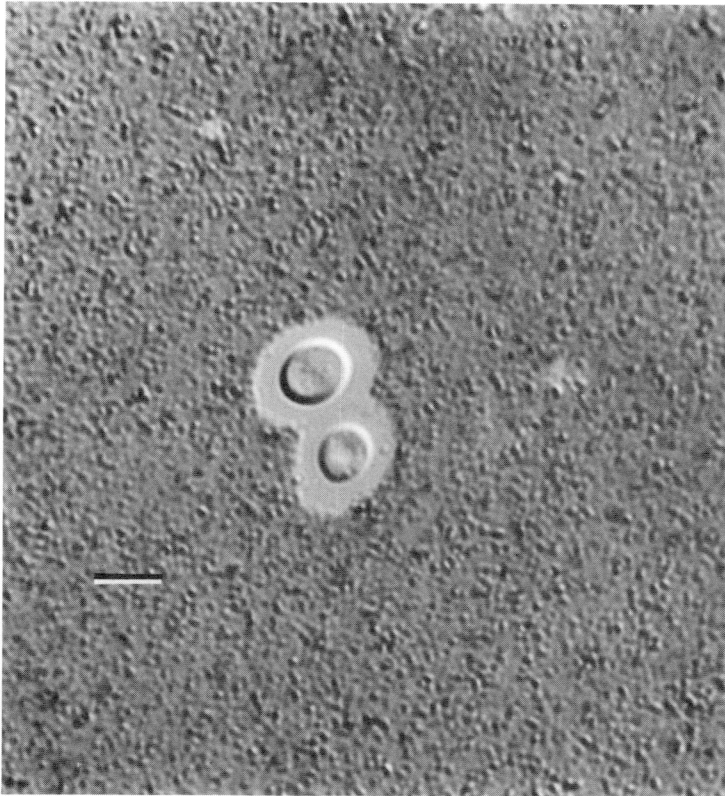

Figure 93. *Cryptococcus laurentii.* The capsule can be seen around the two cells in the India ink preparation. Bar is 10 μm.

Cryptococcus Kützing emend. Phaff et Spencer, 1969

Description: Colonies are fast growing, soft, glistening to dull, smooth, usually mucoid, and cream to slightly pink or yellowish brown in color. Pseudohyphae are absent or rudimentary. Hyphae are absent. Blastoconidia are 1-celled, typically encapsulated, globose to ovoid, with multipolar budding. Members of the genus are unable to ferment sugars, but do assimilate inositol, and produce urease. Carotenoid pigment production is extremely variable. Sexual spores are absent.

Salient characteristics: The genus *Cryptococcus* forms globose to ovoid cells that typically have a capsule, but produces neither pseudohyphae nor true hyphae. Species have the ability to assimilate inositol, but not to ferment sugars. *Cryptococcus neoformans* produces phenoloxidase that results in a brown to black discoloration of the colony when it is grown on caffeic acid agar or birdseed agar. *Cryptococcus* differs from *Rhodotorula* by assimilating inositol; it differs from *Torulopsis* by assimilating inositol and being urease positive; and from *Candida* by not forming well-developed pseudohyphae.

Laboratory precautions: Handle with care, but special precautions are not necessary.

Key references:
Barnett, Payne, and Yarrow, 1979
Lodder, 1970

131

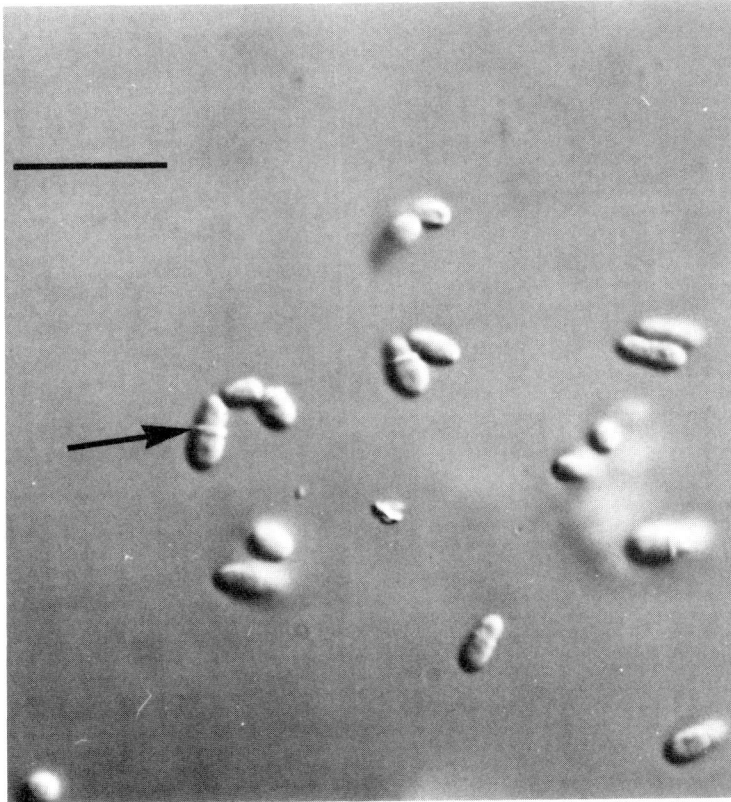

Malassezia Baillon, 1889

Description: Colonies are rapid growing, raised, dull, initially creamy yellow, becoming buff to orange beige (based upon *M. pachydermatis)*. Pseudohyphae and hyphae may be rare in culture. Conidia are 1-celled and globose to ellipsoid. Phialides are flask shaped with a collarette. Fermentation and sexual spores are absent.

Salient characteristics: Isolates of *Malassezia* can be recognized by their bottle shape, unipolar budding, and collarettes at the apices of the phialides.

Laboratory precautions: Handle with care, but special precautions are not necessary.

Key references:
Barnett, Payne, and Yarrow, 1979
Lodder, 1970

Figure 94. *Malassezia pachydermatis.* The yeast cells resemble small bottles. A collarette can be seen at their apical regions (arrow). Bar is 10 μm.

Rhodotorula Harrison, 1927

Description: Colonies are rapid growing, smooth, glistening or dull, sometimes roughened, soft, mucoid, cream to pink or coral red in color. Pseudohyphae are absent or rudimentary. Hyphae are absent. Blastoconidia are 1-celled, globose to elongate, typically with a capsule. Inositol assimilation is negative, fermentation of sugars is absent, and urease is produced.

Salient characteristics: *Rhodotorula* forms pink to coral red colonies having globose to elongate, encapsulated cells, an inability to assimilate inositol or to ferment sugars, and the absence of pseudohyphae. *Rhodotorula* differs from *Cryptococcus* by not being able to assimilate inositol; it differs from *Torulopsis* by producing pink to red colonies and being urease positive; and from *Candida* by not having pseudohyphae.

Laboratory precautions: Handle with care, but special precautions are not necessary.

Key references:
Barnett, Payne, and Yarrow, 1979
Lodder, 1970

133

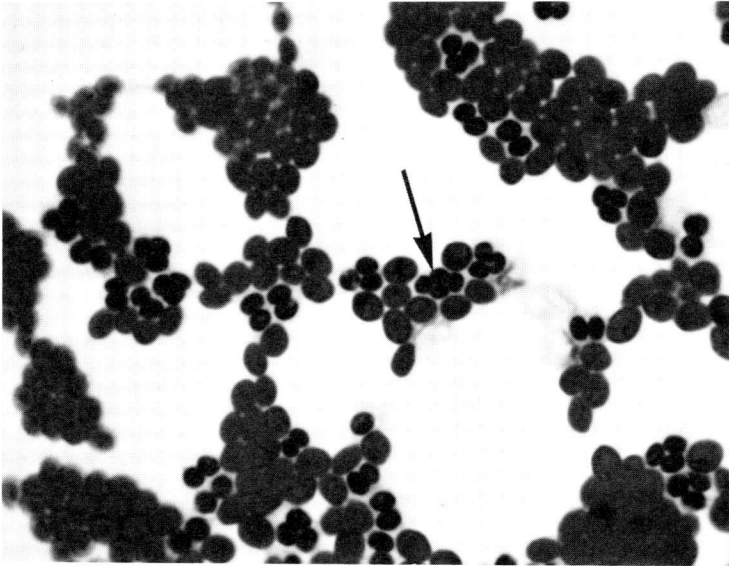

Figure 95. *Saccharomyces cerevisiae.* The ascospores (arrow) are darkly stained. Schaeffer-Fulton modification of the Wirtz stain. (Courtesy of C. Pinello.)

Saccharomyces Meyen ex Hansen, 1883

Description: Colonies are rapid growing, flat, smooth, glistening or dull, and cream to tannish cream in color. Pseudohyphae may be present; when present, they are rudimentary. Hyphae are absent. Blastoconidia are 1-celled, globose, and ellipsoid to elongate. Asci contain 1 to 4 ascospores, and do not rupture at maturity. Ascospores often are globose. Nitrate is not utilized. Fermentation is positive.

Salient characteristics: Typical of the common isolates of *Saccharomyces* are asci containing 1 to 4 ascospores, the inability to utilize nitrate for growth, and 1-celled, multipolar blastoconidia.

Laboratory precautions: Handle with care, but special precautions are not necessary.

Key references:
Barnett, Payne, and Yarrow, 1979
Lodder, 1970

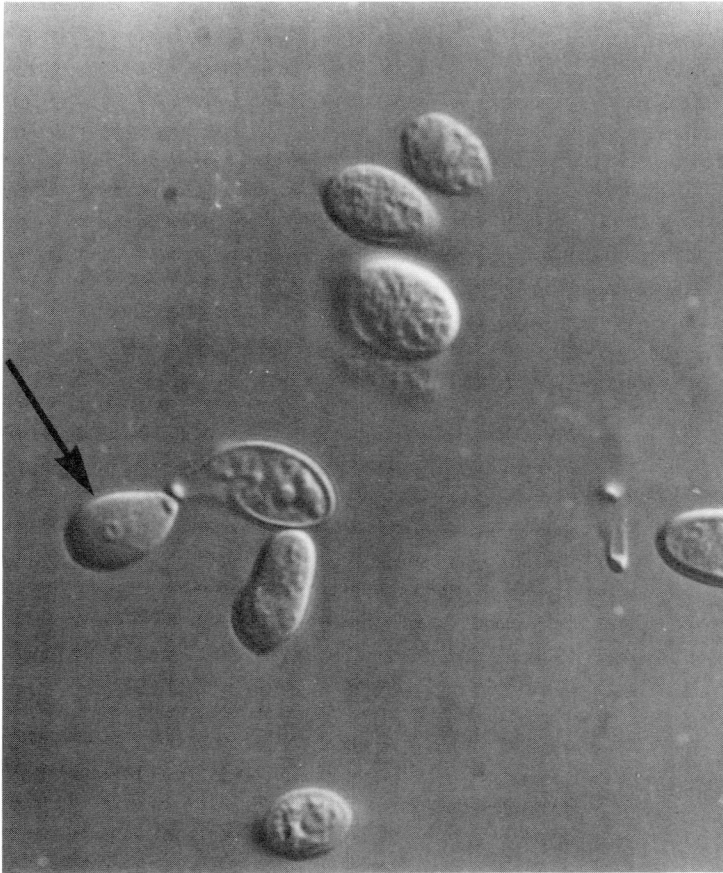

Figure 96. *Sporobolomyces salmonicolor.* A ballistoconidium (arrow) is attached to its parent cell.

Sporobolomyces Kluyver et van Niel, 1924

Description: Colonies are rapid growing, smooth, often wrinkled, glistening to dull, bright red to orange in color. Pseudohyphae and hyphae often are abundant and well developed. Ballistoconidia are 1-celled, usually reniform, forcibly discharged from denticles on ovoid to elongate vegetative cells. Blastoconidia are the most common type of conidia. Fermentation is absent.

Salient characteristics: *Sporobolomyces* is recognized by its development of red to orange colonies and by its typically reniform ballistoconidia, which may be abundant enough to form a mirror image of the colony on nutrient agar on the lid of an inverted Petri plate.

Laboratory precautions: Handle with care, but special precautions are not necessary.

Key references:
Barnett, Payne, and Yarrow, 1979
Lodder, 1970

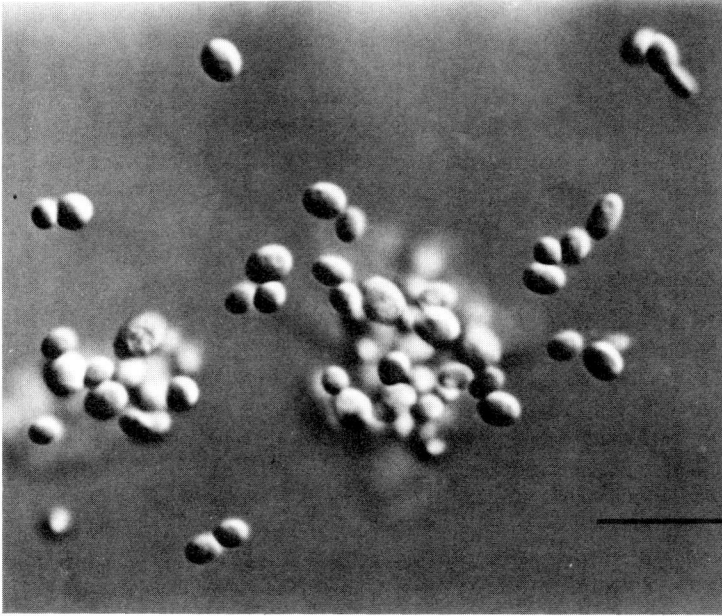

Figure 97. *Torulopsis glabrata.* The yeast cells are globose to ovoid. Bar is 10 μm.

Torulopsis Berlese, 1894

Description: Colonies are rapid growing, glistening, smooth, and cream to white in color. Pseudohyphae are absent or rudimentary. Hyphae are absent. The yeast is 1-celled, globose to ovoid, with multipolar budding. Sexual spores are absent. Fermentation may occur in some species, but inositol assimilation and urease production are negative.

Salient characteristics: Inositol and urease negative yeast isolates not producing pseudohyphae are typical of *Torulopsis*. *Torulopsis* differs from *Candida* by not having well-developed pseudohyphae; it differs from *Cryptococcus* by being unable to grow with inositol as the sole carbon source; from *Rhodotorula* by being urease negative; and from *Saccharomyces* by lacking ascospores.

Laboratory precautions: Handle with care, but special precautions are not necessary.

Key references:
Barnett, Payne, and Yarrow, 1979
Lodder, 1970

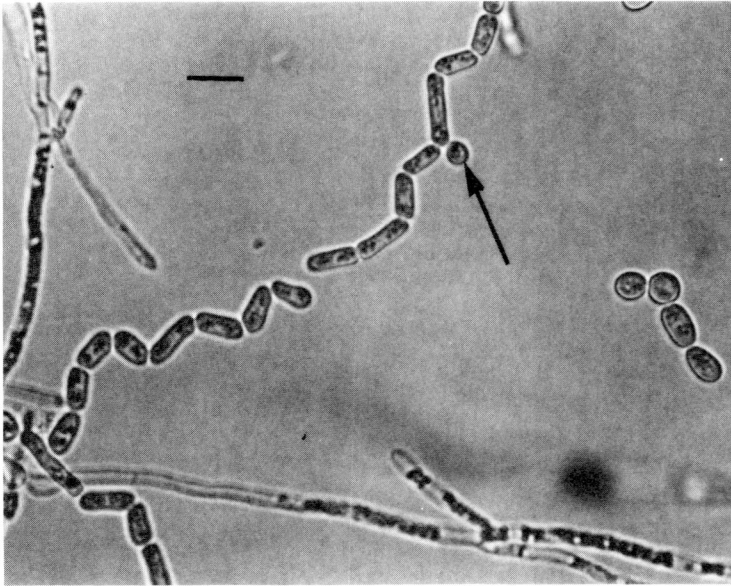

Figure 98. *Trichosporon beigelii*. A blastoconidium (arrow) is developing from an arthroconidium. Bar is 10 μm.

Trichosporon Behrend, 1890

Description: Colonies are rapid growing, smooth, wrinkled, raised, folded, glabrous to velvety, dull, brittle, waxy, white, or yellowish white to cream colored. Pseudohyphae and hyphae are abundant and well developed. Blastoconidia are 1-celled and variable in shape. Arthroconidia are 1-celled, always present, usually elongate. Sexual spores are absent. Fermentation is absent or weak.

Salient characteristics: Isolates of *Trichosporon* produce cream-colored, waxy, raised colonies having 1-celled blastoconidia, and hyphae that fragment into arthroconidia. *Trichosporon beigelii*, the species most commonly encountered in the clinical laboratory, is urease positive. *Trichosporon* differs from *Geotrichum* by having blastoconidia; it differs from *Candida* by forming arthroconidia.

Laboratory precautions: Handle with care, but special precautions are not necessary.

Key references:
Barnett, Payne, and Yarrow, 1979
Lodder, 1970

137

References

Ainsworth, G. C. and K. Sampson. 1950. *The British Smut Fungi (Usti-laginales)*. Commonwealth Mycological Institute, Kew. 137 pp.

Ames, L. M. 1961. *A Monograph of the Chaetomiaceae*. Bibliotheca Mycologica 17, reprinted 1969. J. Cramer, Lehre. 65 pp.

Barnett, J. A., R. W. Payne, and D. Yarrow. 1979. *A Guide to Identifying and Classifying Yeasts*. Cambridge University Press, New York. 315 pp.

Barron, G. L. 1968. *The Genera of Hyphomycetes from Soil*. Williams and Wilkins, Baltimore. 364 pp.

Booth, C. 1971. *The Genus Fusarium*. Commonwealth Mycological Institute, Kew. 237 pp.

———. 1977. *Fusarium Laboratory Guide to the Identification of the Major Species*. Commonwealth Mycological Institute, Kew. 58 pp.

Brown, A. H. S. and G. Smith. 1957. The genus *Paecilomyces* Bainier and its perfect stage *Byssochlamys* Westling. *Trans. Br. Mycol. Soc.* 40:17–89

Carmichael, J. W. 1962. *Chrysosporium* and some other aleuriosporic hyphomycetes. *Can. J. Bot.* 40:1137–73.

Dvorák, J. and M. Otcenásek. 1969. *Mycological Diagnosis of Animal Dermatophytoses*. Academia, Prague. 213 pp.

El-Ani, A. S. 1966. A new species of *Leptosphaeria*, an etiologic agent of mycetoma. *Mycologia* 58:406 11.

——— and M. A. Gordon. 1965. The ascospore sheath and taxonomy of *Leptosphaeria senegalensis*. *Mycologia* 57:275–78.

Ellis, M. B. 1971. *Dematiaceous Hyphomycetes*. Commonwealth Mycological Institute, Kew. 608 pp.

———. *More Dematiaceous Hyphomycetes*. Commonwealth Mycological Institute, Kew. 507 pp.

Fuller, M. S. (ed.). 1978. *Lower Fungi in the Laboratory*. Palfrey Contribut. Bot. No. 1, University of Georgia, Athens. 213 pp.

Gams, W. 1971. *Cephalosporium — artige Schimmelpilze (Hyphomycetes)*. Gustav Fischer, Stuttgart. 262 pp.

Goodfellow, M. and D. E. Minnikin. 1977. Nocardioform bacteria. *Annu. Rev. Microbiol.* 31:159–80.

Gueho, E. and J. Buissière. 1975. Méthode d'identification biochimique de champignons filamenteux arthrosporés appartenant au genre *Geotrichum* Link ex Pers. *Ann. Microbiol.* 126A:483–500.

Hawksworth, D. L. 1979. Ascospore sculpturing and generic concepts in the Testudinaceae (syn. Zopfiaceae). *Can. J. Bot.* 57:91–99.

Hermanides-Nijhof, E. J. 1977. *Aureobasidium* and allied genera. *Stud. Mycol.* 15:141–77.

Hoog, G. S. de. 1972. The genera *Beauveria*, *Isaria*, *Tritirachium* and *Acrodontium* gen. nov. *Stud. Mycol.* 1:1–41.

———. 1974. The genera *Blastobotrys*, *Sporothrix*, *Calcarisporium* and *Calcarisporiella* gen. nov. *Stud. Mycol.* 7:1–84.

———. 1977. *Rhinocladiella* and allied genera. *Stud. Mycol.* 15:1–140.

Huppert, M., S. H. Sun, and E. H. Rice. 1978. Specificity of exoantigens for identifying cultures of *Coccidioides immitis*. *J. Clin. Microbiol.* 8:346–48.

Inui, T., Y. Takeda, and H. Iizuka. 1965. Taxonomical studies on genus *Rhizopus*. *J. Gen. Appl. Microbiol. Suppl.* 11:1–121.

Joly, P. 1964. Le genre *Alternaria*. Recherches Physiologiques, Biologiques et Systematiques. *Encycl. Mycol.* 33:1–250.

Kurylowicz, W. et al. 1975. *Numerical Taxonomy of Streptomycetes.* Polish Medical Publishers, Warsaw. 108 pp.

Lodder, J. (ed.). 1970. *The Yeasts: A Taxonomic Study.* 2d. ed. North-Holland Publishing, Amsterdam. 1,385 pp.

McGinnis, M. R. 1978. Human pathogenic species of *Exophiala, Phialophora,* and *Wangiella. Proc. Int. Conf. Mycoses, 4th.* PAHO Sci. Publ. No. 356, pp. 37–59.

_____. 1980. *Laboratory Handbook of Medical Mycology.* Academic Press, New York. 661 pp.

_____ and D. Borelli, 1981. *Cladosporium bantianum* and its synonym *Cladosporium trichoides.* Mycotaxon 13:127–136.

_____, A. A. Padhye, and L.Ajello. 1982 . *Pseudallescheria* Negroni et Fischer 1943 and its later synonym *Petriellidium* Malloch 1970. Mycotaxon (in press).

Meyer, J. 1976. *Nocardiopsis,* a new genus of the order Actinomycetales. *Int. J. Syst. Bacteriol.* 26:487–93.

Mishra, S. K., R. E. Gordon, and D. A. Barnett. 1980. Identification of nocardiae and streptomycetes of medical importance. *J. Clin. Microbiol.* 11:728–36.

Morton, F. J. and G. Smith. 1963. The genera *Scopulariopsis* Bainier, *Microascus* Zukal, and *Doratomyces* Corda. *Mycol. Paper* 86:1–96.

Oorschot, C. A. N. van. 1980. A revision of *Chrysosporium* and allied genera. *Stud. Mycol.* 20:1–89.

O'Donnell, K. L. 1979. *Zygomycetes in Culture.* Palfrey Contrib. Bot. No. 2, University of Georgia, Athens. 257 pp.

Raper, K. B. and D. I. Fennell. 1965. *The Genus Aspergillus.* Williams and Wilkins, Baltimore. 686 pp.

_____ and C. Thom. 1949. *A Manual of the Penicillia.* Williams and Wilkins, Baltimore. 875 pp.

Rebell, G. and D. Taplin. 1970. *Dermatophytes: Their Recognition and Identification.* University of Miami Press, Coral Gables. 124 pp.

Rifai, M. A. 1969. A revision of the genus *Trichoderma. Mycol. Paper* 116:1–56.

_____ and R. C. Cooke. 1966. Studies on some didymosporous genera of nematode-trapping hyphomycetes. *Trans. Br. Mycol. Soc.* 49:147–68.

Samson, R. A. 1974. *Paecilomyces* and some allied hyphomycetes. *Stud. Mycol.* 6:1–119.

_____. 1979. A compilation of the aspergilli described since 1965. *Stud. Mycol.* 18:1–40.

_____, R. Hadlok, and A. C. Stolk. 1977. A taxonomic study of the *Penicillium chrysogenum* series. *Antonie van Leeuwenhoek J. Microbiol. Serol.* 43:169–75.

_____, A. C. Stolk, and R. Hadlok. 1976. Revision of the subsection fasciculata of *Penicillium* and some allied species. *Stud. Mycol.* 11:1–47.

Schipper, M. A. A. 1978a. 1. On certain species of *Mucor* with a key to all accepted species. *Stud. Mycol.* 17:1–52.

_____. 1978b. 2. On the genera *Rhizomucor* and *Parasitella. Stud. Mycol.* 17:53–71.

Schol-Schwarz, B. 1959. The genus *Epicoccum* Link. *Trans. Br. Mycol. Soc.* 42:149–73.

Shirling, E. B. and D. Gottlieb. 1972. Cooperative description of type strains of *Streptomyces. Int. J. Syst. Bacteriol.* 22:265–394.

Sigler, L. and J. W. Carmichael. 1976. Taxonomy of *Malbranchea* and some other hyphomycetes with arthroconidia. *Mycotaxon* 4:349–488.

Simmons, E. G. 1967. Typification of *Alternaria, Stemphylium,* and *Ulocladium. Mycologia* 59:67–92.

Somal, B. S. 1976. A key to the species of *Curvularia. Indian J. Mycol. Plant Pathol.* 6:59–64.

Sutton, B. C. 1980. *The Coelomycetes.* Commonwealth Mycological Institute, Kew. 696 pp.

Sykes, G. and F. A. Skinner (eds). 1973. *Actinomycetales: Characteristics and Practical Importance.* Academic Press, New York. 339 pp.

Takashio, M. and R. Vanbreuseghem. 1971. Production of ascospores by *Piedraia hortai in vitro. Mycologia* 63:612–18.

Upadhyay, H. P. and W. B. Kendrick. 1974. A new *Graphium*-like genus (conidial state of *Ceratocystis). Mycologia* 66:181–83.

Vries, G. A. de. 1952. *Contribution to the knowledge of the genus Cladosporium Link ex Fr.* Uitgeverij and Drukkerij Hollandia Press, Baarn. 121 pp.

Weijman, A. C. M. 1979. Carbohydrate composition and taxonomy of *Geotrichum, Trichosporon* and allied genera. *Antonie van Leeuwenhoek J. Microbiol. Serol.* 45:119–27.

Weitzman, I. and M. Y. Crist. 1979. Studies with clinical isolates of *Cunninghamella.* I. Mating behavior. *Mycologia* 71:1024–33.

Zycha, H., R. Siepmann, and G. Linnemann. 1969. *Mucorales: Eine Beschreibung aller Gattungen und Arten dieser Pilzgruppe.* J. Cramer, Lehre. 355 pp.

Glossary

Acid-fast. A property of cell walls that during a staining reaction retain basic dyes when decolorized with mineral acids.

Acropetal. Having the youngest conidia at the apex of a chain.

Actinomycete. A Gram-positive bacterium that grows vegetatively in a branched filamentous form.

Aerial hyphae. Hyphae that grow above the agar surface.

Aerobic. Having the ability to grow in the presence of oxygen.

Allantoid. Sausage shaped.

Anaerobic. Lacking the ability to grow in the presence of oxygen.

Annellide. A conidiogenous cell that gives rise to successive conidia in a basipetal manner. The apex of an annellide becomes longer and narrower as each subsequent conidium is formed and released. An apical ring composed of outer cell wall material, remains as each conidium is released.

Annelloconidium (pl. annelloconidia). A conidium formed by an annellide.

Annular frill. A ring or skirtlike portion of cell wall material at the base of a conidium that remains when the conidium separates from its conidiophore.

Antler hypha. See favic chandelier.

Apex (pl. apices). The tip.

Arthroconidium (pl. arthroconidia). A conidium formed by the modification of a hyphal cell(s) and then released by the fragmentation-lysis of a disjunctor cell or by fission through a thickened septum.

Arthrospore. See arthroconidium.

Ascospore. A haploid sexual spore that is formed by free-cell formation in an ascus following karyogamy and meiosis.

Ascostroma (pl. ascostromata). A specialized mass of hyphae containing cavities in which asci develop.

Ascus (pl. asci). A saclike cell that gives rise to ascospores. Asci are characteristic of the Ascomycetes.

Assimilation. The utilization of nutrients for growth, with oxygen serving as the final electron acceptor. $C_6H_{12}O_6 + 6O_2 \rightarrow 6CO_2 + 6H_2O$.

Bacterium (pl. bacteria). A simple prokaryotic microorganism having absorptive nutrition.

Ballistoconidium (pl. ballistoconidia). A forcibly discharged conidium.

Balloon form. Pertaining to a large globose conidium formed by some dermatophytes, especially *Trichophyton tonsurans*.

Basidiospore. A haploid sexual spore formed on a basidium following the process of karyogamy and meiosis.

Basidium (pl. **basidia**). A specialized cell that gives rise to basidiospores. Basidia are characteristic of the Basidiomycetes.

Basipetal. Having the youngest conidia at the base of a chain.

Bipolar budding. The development of conidia at both ends of the parent cell.

Biseriate. Having phialides arising from metulae on the vesicles of species of *Aspergillus.*

Bitunicate. Having two walls.

Black yeast. A dematiaceous, unicellular, budding fungus that typically forms a black, pasty colony.

Blastoconidium (pl. **blastoconidia**). A conidium that is blown out from part of its parent cell and is typically released by fission through a thickened basal septum.

Blastospore. See blastoconidium.

Budding. Asexual formation of small, rounded outgrowths from a parent cell. These will become conidia.

Capsule. A gelatinous covering around a cell.

Carry-over. Indigenous substances stored within the cells of inoculum, nutrients in the original culture medium, or both. These substances support growth of the test isolate in an assimilation study.

Chlamydoconidium (pl. **chlamydoconidia**). A rounded, enlarged conidium that usually has a thickened cell wall and functions as a survival propagule.

Chlamydospore. See chlamydoconidium.

Circinate. Coiled into a complete or partial ring.

Clamp connection. A specialized hyphal bridge involved with nuclear division in the Basidiomycetes.

Clavate. Club shaped.

Cleistothecium (pl. **cleistothecia).** An enclosed fruiting body that contains randomly dispersed asci.

Collarette. A small collar.

Columella (pl. **columellae).** A sterile domelike expansion at the apex of a sporangiophore.

Conical. Cone shaped.

Conidiogenous cell. A cell that gives rise to conidia.

Conidiophore. A specialized hypha upon which conidia develop.

Conidium (pl. **conidia).** An asexual, nonmotile, usually deciduous propagule that is not formed by cytoplasmic cleavage, free-cell formation, or by conjugation.

Coremium (pl. **coremia).** See synnema.

Cottony. See floccose.

Crosswall. A septum.

Dematiaceous. Having brown to black conidia or hyphae.

Denticle. A peg.

Dermatophyte. A fungus in the genus *Epidermophyton, Microsporum,* or *Trichophyton* that infects hair, nail, or skin.

Dimorphic. Having two different morphologic forms.

Disjunctor cell. A cell that releases a conidium by its fragmentation or lysis.

Double septum. A thickened septum that separates through its center to release a conidium.

Echinulate. Having a delicate, spiny wall.

Endospore. A spore formed within a spherule by a cleavage process following karyogamy and mitosis.

Erect. Upright.

Exudate.   Droplets of fluid formed on the surface of a colony.

Favic chandelier.   A repeatedly branched cluster of hyphal apices that resembles a chandelier.

Fermentation.   The ability to utilize nutrients for growth, with organic compounds serving as the final electron acceptors. $C_6H_{12}O_6 \rightarrow 2C_2H_5OH + 2CO_2$.

Filament.   A threadlike element of a bacterium; a hypha of a fungus.

Fission.   To split into two portions or cells.

Fission arthroconidium.   An arthroconidium that is released by fission through a double septum.

Floccose.   Having a cottony texture.

Foot cell.   The base of a macrophialoconidium produced by a species of *Fusarium* having a heellike projection; the base of the conidiophore of *Aspergillus* species where it merges with the hypha and resembles the heel and toes of a foot.

Fragmentation.   Separation of a hypha into conidia.

Fungus (pl. fungi).   A eukaryotic, unicellular to filamentous, achlorophyllous organism having absorptive nutrition. A fungus reproduces by sexual, asexual, or both means.

Fusiform.   Tapering at both ends; spindle shaped.

Geniculate.   Bent like a series of knees.

Germ pore.   An unthickened spot in a spore or conidial wall through which a germ tube may form.

Germ tube.   A hyphal initial developing from a conidium or spore.

Glabrous.   Smooth.

Globose.   Round.

Hemispheric.   Half of a sphere.

Hilum (pl. hila).   A scar at the base of a conidium.

Hulle cell.   A cell having a thickened cell wall and a small lumen. Hulle cells are associated with species of *Aspergillus*.

Hyaline.   Without color.

Hypha (pl. hyphae).   An individual filament of a fungus.

Intercalary.   Occurring within a hypha.

Internode.   That portion of a hypha that is between two nodes.

Karyogamy.   Fusion of two nuclei.

Lageniform.   Flask shaped with a tapering distal portion.

Lanose.   Having a woolly texture.

Lysis.   Dissolution.

Macroconidium (pl. macroconidia).   The larger of two conidia of two different sizes that are produced in the same manner by a single fungus.

Merosporangium (pl. merosporangia).   A sporangium having its sporangiospores in a single row.

Metula (pl. metulae).   A sterile branch upon which phialides of some species of *Aspergillus* and *Penicillium* develop.

Microconidium (pl. microconidia)   The smaller of two conidia of two different sizes that are produced in the same manner by a single fungus.

Moniliform.   Having swellings.

Mould.   A filamentous fungus.

Multiple budding.   The development of several series of blastoconidia around a parent yeast cell.

Muriform.   Having vertical and horizontal septa.

Mycelium.   The aggregated mass of hyphae making up a fungus.

Mycology.   The branch of biology that deals with the study of fungi.

Node.   Where a stolon touches a surface.

Nodular organ.   A knot of hyphae that is often produced by dermatophytes.

Obclavate.   Club shaped in reverse.

Obovoid. Egg shaped in reverse.

Olivaceous. Having an olive shade of color.

Ostiolate. Having an ostiole.

Ostiole. A mouth or opening through which spores or conidia may escape.

Oval. Egg-shaped.

Papilla (pl. **papillae**). A small nipple-shaped elevation.

Penicillus. A brushlike conidial head produced by members of the genus *Penicillium*.

Percurrent. Developing through a previous apex.

Perforating organ. A mass of hyphae producing a conical cavity in the *in vitro* hair test; a characteristic property of certain keratinophilic fungi.

Perithecium (pl. **perithecia**). A fruiting body having asci in a basal group or as a layer. Perithecia are usually flask shaped, with an opening through which the asci or ascospores escape.

Phialide. A type of conidiogenous cell that gives rise to successive conidia from a fixed site in a basipetal manner. A phialide does not increase in length as the conidia are formed, and its apex does not become smaller in diameter. A collarette is often present at the apex of the phialide.

Phialoconidium (pl. **phialoconidia**). A conidium produced by a phialide.

Phycomycetes. An archaic class name once used for the lower fungi in general. These organisms are now placed either in the kingdom Protista or in the classes Trichomycetes and Zygomycetes of the kingdom Fungi.

Pleomorphic. Having several forms. The term is also applied to dermatophyte colonies that become irreversibly sterile.

Polymorphic. Having several forms.

Poroconidium (pl. **poroconidia**). A conidium that forms through a pore in the cell wall of its conidiogenous cell.

Propagule. A reproductive unit.

Pseudohypha (pl. **pseudohyphae**). A series of blastoconidia that have remained attached to each other forming a filament. The blastoconidia are often elongated with the points of attachment between adjacent cells being constricted.

Pseudomycelium. A large amount of pseudohyphae.

Pycnidium (pl. **pycnidia**). A saclike fruiting body that gives rise to conidia within its central area.

Pyriform. Pear shaped.

Rachis. An extension of a conidiogenous cell-bearing conidia.

Racket hypha (also spelled **racquet**). A hypha having a series of cells that are swollen at one end.

Radiating. Spreading from a common center.

Reflexive hypha. A hypha having short branches that bend backward at approximately a 45°angle.

Rhizoid. Pertaining to a rootlike group of hyphae.

Rudimentary. Poorly developed.

Sclerotium (pl. **sclerotia**). An organized mass of hyphae that remains dormant during unfavorable conditions.

Septum (pl. **septa**). A crosswall.

Seta (pl. **setae**). A bristle or bristlelike structure.

Shield cell. A conidium having the shape of a shield. Shield cells are commonly produced by members of the genus *Cladosporium*.

Simple. Of one piece; unbranched.

Sinuous. Wavy.

Solitary. Separate; alone.

Spherule. A sporangiumlike structure containing endospores that is produced by *Coccidioides immitis* or *Rhinosporidium seeberi.*

Sporangiolum (pl. **sporangiola**). A sporangium that contains a small number of sporangiospores. Some sporangiola may contain only one sporangiospore.

Sporangiophore. A specialized hypha that gives rise to a sporangium.

Sporangiospore. A spore that is formed by a cleavage process following karyogamy and mitosis in a sporangium.

Sporangium (pl. **sporangia**). An asexual saclike cell that has its entire content cleaved into sporangiospores.

Spore. A reproductive propagule that forms either following meiosis or asexually by a cleavage process.

Sporodochium (pl. **sporodochia**). A cushionlike mat of hyphae bearing conidiophores over its surface.

Sterigma (pl. **sterigmata**). A pedicel bearing a basidiospore.

Stolon. A runner.

Subglobose. Almost round.

Submerged. Within the nutrient agar.

Sympodial. Pertaining to the growth of a conidiophore in which new successive lateral, subterminal apices of growth occur following successive conidium formation. Sympodial conidiophores are typically geniculate in appearance.

Synnema (pl. **synnemata**). An erect macroscopic structure consisting of united conidiophores that bear conidia terminally, laterally, or in both ways.

Truncate. Ending abruptly.

Tuberculate. Having fingerlike or wartlike projections.

Unipolar budding. The development of conidia at one end of the parent cell.

Uniseriate. Having phialides that arise directly from the vesicle in species of *Aspergillus.*

Verrucose. Having warts.

Verticil. A whorl of conidiogenous cells or conidiophores arising from a common point.

Verticillate. Having verticils.

Vesicle. A swollen cell; the swollen apices of some conidiophores or sporangiophores.

Villose. Bearing long, hairlike appendages.

Yeast. A unicellular budding fungus that reprodues by sexual, asexual, or both means.

Yeastlike. Pertaining to a unicellular budding fungus that reproduces by asexual means only.

Zonate. Having concentric bands of color or growth.

Zygospore. A resting spore in which meiosis will occur. Zygospores result from the fusion of two similar hyphal elements. They are characteristic of the Zygomycetes.

Materials and Methods for Identification

APPENDIX A

All media and reagents marked with an asterisk are readily available commercially. The methods of preparation for media and reagents not commercially available are included in Appendix B.

I. Aerobic Actinomycetes
   A. Casein, Xanthine, Hypoxanthine, and Tyrosine Hydrolysis
      1. With a marking pen, divide the Petri plates containing each of the above substrates into 4 quadrants.
      2. Make a point inoculation of approximately 5 mm in the center of each quadrant for each isolate.
      3. Seal the plates with parafilm and incubate at 25–30°C for the following durations: casein, 2 weeks; xanthine, 3 weeks; hypoxanthine, 3 weeks; tyrosine, 4 weeks. Because some tests may become positive more rapidly at 35°C, a duplicate set of media may be incubated at this temperature.
      4. Examine plates at intervals of 3 to 4 days for clearing of the medium underneath and around the growth of the organism.
      5. The following control organisms should be set up with each test:
         positive control, *Streptomyces griseus*
         negative control, *Nocardia asteroides.*
   B. Cellobiose, Acid from
      1. Inoculate a slant of cellobiose medium with 0.1 ml of a suspension of the organism to be identified.
      2. Incubate the tube at 25–30°C for 4 weeks.
      3. Examine at intervals of 3 to 4 days for evidence of acid production (during which the medium will change from purple to yellow).
      4. The following controls should be set up with each test:
         positive control (yellow), *Actinomadura madurae*
         negative control (purple), *Nocardia asteroides.*
   C. Lysozyme Resistance
      This test is more reliable than the acid-fast technique in the laboratory for the identification of the most commonly recovered isolates of *Nocardia. Streptomyces griseus* should be used as the negative control because it is consistently negative.
      1. Place several pieces of the isolate to be tested in a tube containing glycerol broth with lysozyme (control).
      2. Incubate at 25–30°C until the control tube shows evidence of good growth (up to one week).
      3. If the isolate grows well in both tubes, it is considered to be resistant to lysozyme. If good growth is present in the control tube and, if poor or no growth in the tube containing lysozyme, the isolate is susceptible to lysozyme.
      4. The following control organisms should be set up with each test:
         positive control, *N. asteroides*
         negative control, *S. griseus.*
   D. Slide Culture
      1. Sterilize a 15 × 100 mm glass Petri plate containing a piece of filter paper, upon which is placed a bent (V-shaped) glass rod, a glass microscope slide, and a No. 1 coverslip. Sterilize in a hot air oven or autoclave.

2. Aseptically place the glass slide on the top of the bent glass rod and transfer a 1-cm square agar block of either brain-heart infusion agar,* potato dextrose agar,* or Sabouraud dextrose agar* to the center of the slide. The agar block should be taken from a plate that contains approximately 35 ml of medium.

3. Using a sterile inoculating needle, inoculate each side of the agar block with small pieces of the isolate.

4. With a flamed forceps, place the coverslip on the agar block.

5. Moisten the filter paper with approximately 1 ml of sterile water. The filter paper must remain moist throughout the incubation period.

6. Place the cover on the Petri plate and incubate at 30–35°C.

7. Examine the slide culture microscopically every other day by removing the slide from the Petri plate, wiping its bottom, and then placing it on the stage of a compound microscope.

E. Urea Hydrolysis

1. Transfer a large amount of the organism to be tested to a tube of urea broth.*

2. Incubate at 25–30°C for 4 weeks.

3. Examine the tubes at intervals of 3 to 4 days. Hydrolysis is seen as an alkaline reaction, during which the medium turns from yellow to red.

4. The following control organisms should be set up with each test run:
positive control (red), *N. brasiliensis*
negative control (yellow), *A. madurae*.

II. Moulds

A. Hair Perforation, *In vitro*

This test is used for the identification of various dermatophytes, especially for differentiating *Trichophyton mentagrophytes* from *T. rubrum*. The former penetrates hair perpendicularly, resulting in conical or wedge-like perforations; the latter does not.

1. Sterilize healthy human hair fragments by autoclaving at 121°C for 15 minutes. Blond hair from a child is ideal.

2. Place several fragments in a sterile Petri plate.

3. Add 20 ml of sterile distilled water and 0.1 ml of 10% sterile yeast extract.

4. Transfer a few fragments of the dermatophyte to be tested to the Petri plate.

5. Incubate at 25–30°C and examine weekly for up to 4 weeks. Prepare a lactophenol cotton blue or lactophenol mount of a hair fragment and observe microscopically.

6. The following control organisms should be set up with each test:
positive control, *T. mentagrophytes*
negative control, *T. rubrum*.

B. Mould-Yeast Conversion, *In vitro*

1. Transfer a small portion of the fungus to be tested to 2 tubes of conversion medium (see Table A.1).

2. Incubate one tube at 35°C and the other at 25°C.

3. Prepare lactophenol cotton blue or lactophenol mounts and observe microscopically weekly for the appropriate yeast morphology, the conversion of hyphae to the characteristic yeast form, or both. The entire culture does not have to convert to the corresponding yeast form. *These preparations should be made in a biological safety cabinet.*

C. Nutritional Tests for *Trichophyton* Species
Nutritional tests are used to confirm the identity of species of *Trichophyton* (especially those species that resemble one another), and to identify isolates that are not producing conidia.

TABLE A.1   Media and Incubation Conditions for *In Vitro* Mould-Yeast Conversion[a]

Organism	Medium	Incubation
Blastomyces dermatitidis	BHIA[b]	35°C
Histoplasma capsulatum	BHIA + 5% blood	35°C
Paracoccidioides brasiliensis	BHIA	35°C
Sporothrix schenckii	BHIA	35°C + 5–10% CO_2

[a]Confirmation of *Coccidioides immitis* is done with exoantigens. Modified Converse medium can be used for mould-spherule conversions.
[b]Brain-heart infusion agar.*
Source: Modified from McGinnis. 1980. *Laboratory Handbook of Medical Mycology.* Academic Press.

1. Transfer a small portion (≤1 mm) of the dermatophyte to be tested to the center of the slant of the following nutritional media:
   Casein basal medium (medium No. 1)*
   Casein basal medium + inositol (medium No. 2)*
   Casein basal medium + thiamine + inositol (medium No. 3)*
   Casein basal medium + thiamine (medium No. 4)*
   Casein basal medium + nicotinic acid (medium No. 5)*
   Casein basal medium + L-histidine (medium No. 7).*
   Avoid transferring agar with the inoculum, in order to eliminate carry-over.
2. Incubate the media at 25-30°C and examine weekly for 4 weeks. Media used to confirm the identity of *T. verrucosum* are incubated at 35°C for 2 weeks.
3. Test results are recorded as 0-4+. Zero represents no growth; 4+ represents the greatest amount of growth.

D. Rice-Grain Test

This test is used to confirm the identity of *Microsporum audouinii*. *Microsporum audouinii* will not grow on unfortified polished white rice grains although it may produce a brownish discoloration under the colony, whereas *M. canis* and other fungi will grow on this substrate.

1. Transfer a few fragments of the dermatophyte to be tested to a flask containing sterile rice grains.
2. Incubate at 25-30°C and examine for growth at 8 to 10 days.

E. Slide Culture (Set-up)

Slide cultures are prepared when the identification of a mould cannot be made with a teased preparation because it is difficult to determine how the conidia were produced.

1. Prepare a Petri plate containing potato dextrose agar* or cornmeal agar.* Each plate should contain approximately 35 ml of medium.
2. Set up the slide culture as outlined in the aerobic actinomycete section.
3. Incubate at 25-30°C in the darkness.

F. Slide Culture (Examination)

1. Examine the slide culture as outlined in the aerobic actinomycete section.
2. When the slide culture has reached the stage of development that permits identification of the mould, gently remove the coverslip with forceps.

3. Pass the coverslip rapidly through an open flame.
4. Place the edge of the coverslip next to a small drop of lactophenol cotton blue or lactophenol mounting medium on a clean glass microscope slide. Gently lower the coverslip onto the mounting medium, avoiding the entrapment of air bubbles.
5. Observe microscopically.
6. Permanent mounts are prepared by removing excess mounting medium with a paper towel and then applying fingernail polish around the edges of the coverslip.

G. Tease Mount

1. Place a drop of lactophenol cotton blue or lactophenol mounting medium on a microscope slide.
2. With a sterile probe, remove a fragment of the colony to be examined. Select material midway between the center and the edge of the colony.
3. Place the fragment into the drop of mounting medium and tease gently with two dissecting needles.
4. Place the edge of a coverslip next to the drop of the mounting medium and gently lower the coverslip onto the mounting medium, avoiding the entrapment of air bubbles. Do not tap or press the coverslip down.
5. Observe microscopically.
6. For the preparation of permanent mounts, ring the coverslip with fingernail polish after the excess mounting fluid has been removed.

H. Temperature Tolerance Test

1. Transfer a small portion of the fungus to be tested to two Sabouraud dextrose agar* slants.
2. Incubate one tube at 25-30°C and the other at the desired elevated temperature.
3. Growth at both temperatures indicates a positive test; growth at only the lower temperature indicates a negative test. If an organism fails to grow at the lower temperature, the test must be repeated.

I. Urea Hydrolysis

Detection of urease activity within 5 days aids in the differentiation of *Trichophyton mentagrophytes* (positive) from *T. rubrum* (negative). It must be remembered that some isolates of *T. rubrum* may be urease positive, whereas some isolates of *T. mentagrophytes* may be negative. The test must be read within 5 days.

1. Transfer a small portion of the dermato-

phyte to be tested to a slant of Christensen's urea agar.*

2. Incubate at 25–30°C for 5 days. Hydrolysis is evidenced by an alkaline reaction, during which the medium turns from yellow to red.

3. The following control organisms should be set up with each test:
positive control (red), *T. mentagrophytes*
negative control (yellow), *T. rubrum*.

III. Yeasts

A. Ascospore Production

1. Transfer a portion of the yeast to be tested to V-8 juice agar or Wickerham's yeast-malt agar.

2. Incubate at 25–30°C for five days and examine for the presence of ascospores according to one of the methods below. If ascospores are not observed, reincubate the medium for an additional seven days.

a. Schaeffer-Fulton Modified Wirtz Stain

1) Heat-fix a smear of the growth.
2) Flood the smear with 5% aqueous malachite green for 60–90 seconds.
3) Heat to steaming three to four times.
4) Wash with running water for one minute.
5) Counterstain with 0.5% safranin for 30 seconds.
6) Ascospores stain blue green, and vegetive cells stain red.

b. Kufferath-Carbolfuchsin Stain

1) Heat-fix a smear of the growth.
2) Flood the smear with Ziehl-Neelsen carbolfuchsin.
3) Steam for 2.5 minutes.
4) Decolorize with acid alcohol or with 2% lactic acid.
5) Counterstain with methylene blue for one minute.
6) Ascospores stain red, and vegetive cells stain blue.

3. The following control organisms should be set up with each staining procedure:
positive control, *Saccharomyces cerevisiae*
negative control, a non-ascomycetous yeast (*Torulopsis glabrata,* etc.)

B. Carbon Assimilation Test (Auxanographic Method)

1. Using sterile distilled water, prepare a yeast suspension from a pure 24 to 48 hour culture equal to a MacFarland #5 turbidi-

ty standard. *Do not transfer growth medium.*

2. Add 1 ml of yeast suspension to 25 ml of yeast nitrogen base* that has been melted and cooled to 50°C; then mix carefully.

3. Pour the resulting suspension into a sterile 15 × 150-mm plastic Petri plate and allow the medium to solidify.

4. On the agar surface, place paper discs containing the carbon sources* to be tested. Do not place more than 7 discs on a plate.

5. Incubate the plates, agar surface up, at 25–30°C.

6. Examine the plates daily for 4 days. Growth around a disc represents a positive reaction. If no growth is present around the glucose disc, repeat the test.

C. Dalmau Plate

1. With a marking pen, divide in half a cornmeal-Tween 80* agar plate.

2. With a sterile inoculating needle, lightly touch the yeast colony to be tested.

3. Make 2 streaks on the surface of the medium approximately 3 cm long and 1 cm apart. Streak back and forth several times across the 2 streak lines. Do not dig into the surface of the medium.

4. Flame sterilize a 22-mm coverslip, allow it to cool, and place it over the streak marks.

5. Incubate the plate at 25–30°C.

6. Examine the plate daily for 3 days by removing the cover of the Petri plate and placing the bottom plate (agar surface up) on the microscope stage. Using the low power objective, bring the edge of the coverslip into focus and observe the growth. For more detailed study, the high dry objective may be used.

7. A strain of *Candida albicans* known to produce chlamydospores should be set up with each test. If the control strain does not produce chlamydospores, the test must be repeated.

D. Germ-Tube Test
Approximately 95% of *Candida albicans* strains and some strains of *C. stellatoidea* develop germ tubes.

1. Add 0.5 ml of sterile bovine serum or 2% peptone broth to a test tube.

2. With a clean Pasteur pipette, lightly touch the yeast colony to be tested and suspend the yeast cells in the serum, leaving the pipette in the tube. Use a light inoculum.

3. Incubate at 35°C for 2.5–3.0 hours.

4. Place a drop of serum on a clean glass microscope slide and cover with a coverslip.

5. Observe microscopically for the presence of germ tubes.
6. The following control organisms should be set up with each test:
positive control, *C. albicans*
negative control, *C. tropicalis.*

E. Nitrate Assimilation Test
1. Auxanographic method
   a. Inoculum is prepared in the same manner as for the carbon assimilation test.
   b. Add 1 ml of yeast suspension to 15 ml of yeast carbon base* that has been melted and cooled to 50°C; then mix carefully.
   c. Pour the resulting suspension into a sterile 15 × 100-mm plastic Petri plate and allow the medium to solidify.
   d. Divide the Petri plate in half and label one side potassium nitrate (KNO$_3$) and the other side peptone.
   e. Place paper discs containing KNO$_3$ and peptone on the agar surface.
   f. Incubate the plates, agar surface up, at 25–30°C.
   g. Examine the plates daily for 4 days. Growth around both discs indicates a positive nitrate assimilation test. Growth around the peptone disc only indicates a negative test. If no growth is present around the peptone disc, repeat the test.
   h. The following control organisms should be set up with each test:
   positive control, *Cryptococcus albidus*
   negative control, *C. neoformans.*
2. Swab-Nitrate Test
   a. Saturate cotton-tipped swabs in 5×-strength swab nitrate test reagent.
   b. Dry the swabs for 24 hours at room temperature.
   c. Sweep the swab across the yeast growth to be tested.
   d. Work the yeasts into the swab by swirling the swab against the sides of a test tube.
   e. Incubate the swab in the test tube for 10 minutes at 45°C.
   f. Place 2 drops each of 0.5% α-naphthylamine and 0.8% sulfanilic acid[a] into a second test tube.
   g. Place the swab in the second tube.
   h. The same controls used in the auxanographic method are used in the rapid nitrate test.
   i. Bright red color within 5 minutes indicates a positive reaction for nitrate reduction
   If the reaction is light red or pink, or if the test takes longer than 5 minutes, the reagents are too old; the test should be repeated with fresh reagents.

F. Urea Hydrolysis
1. Method I
   a. Transfer a portion of the yeast to be tested to a Christensen's urea agar* slant.
   b. Incubate at 25–30°C for 5 days and examine for hydrolysis as evidenced by an alkaline reaction (medium turns from yellow to red).
   c. The following control organisms should be set up with each test:
   positive control (red), *Cryptococcus albidus*
   negative control (yellow), *Candida albicans.*
2. Method II
   a. On the day of use, reconstitute a vial of Difco urea R broth* according to the manufacturer's instructions.
   b. Place 0.2 ml of the broth in the well(s) of a microtiter plate.
   c. Transfer a heavy inoculum of yeast to be tested to a well containing broth.
   d. Seal the incubated wells with clear tape, incubate at 35°C for 4 hours, and examine for hydrolysis as evidenced by an alkaline reaction (medium turns from yellow to red).
   e. The same control organisms used in Method I are used in this test.
3. Method III (Rapid Selective Test for *Cryptococcus neoformans*)
   a. Reconstitute Difco urea R broth* as a 5× solution, and adjust the pH to 5.5.
   b. Sterilize the solution by filtration.
   c. Saturate a number of sterile cotton-tipped swabs with the sterile 5× solution and allow them to dry overnight at ambient temperature.

[a]α-naphthylamine can be purchased from Sigma Chemical Company. A pair of alternate reagents are 0.5% p-arsanilic acid in 5N acetic acid and 0.8% n-naphthylethylenediamine dihydrochloride in 5N acetic acid. The color produced is a reddish purple; and on the limited scale thus far used, this pair of reagents appear to be as sensitive as the first pair of reagents.

d. Prepare a 1% solution of Zephiran Chloride and adjust pH to exactly 4.86.

e. Prior to testing, immerse the urea swab in the 1% Zephiran Chloride solution.

f. Sweep a moistened swab across 4 colonies of the unknown yeast.

g. Swirl the swab several times against the sides of a test tube and then incubate at 45°C for 15 minutes.

h. The same controls used in Method I are used in this test.

i. If the swab turns pink within 10 minutes, there is an 85-90% probability that the isolate is *C. neoformans*.

If the swab turns pink after 10 minutes, the yeast is probably not *C. neoformans*.

4. Method IV. (Rapid Selective Test for *Cryptococcus neoformans*)

a. Prepare a 5× solution of urea agar base (sans agar) and then adjust the pH to 5.5

b. Sterilize by filtration.

c. Prepare cotton-tipped swabs and the 1% Zephiran Chloride solution as in Method III.

d. Prior to testing, dip the dried swab into 1% Zephiran Chloride at pH 4.86.

e. The same controls used in Method I are used here.

f. Isolates of *C. neoformans* turn the swabs pink within 15 minutes.

Urease determinations for negative isolates must be made by another method.

References

Haley, L. D., and C. S. Callaway. 1978. *Laboratory Methods in Medical Mycology.* Fourth Ed. HEW Publ. No. (CDC) 78-8361. U.S. Government Printing Office, Washington, D.C. 225 pp.

Hopkins, J. M. and G. A. Land. 1977. Rapid method for determining nitrate utilization by yeasts. *J. Clin. Microbiol.* 5:497-500.

Land, G. A. 1981. Rapid selective urease test for *Cryptococcus neoformans. J. Clin. Microbiol.* (Submitted).

McGinnis, M. R. 1980. *Laboratory Handbook of Medical Mycology.* Academic Press, New York. 661 pp.

Mishra, S. K., R. E. Gordon, and D. A. Barnett. 1980. Identification of nocardiae and streptomycetes of medical importance. *J. Clin. Microbiol.* 11:728-36.

Philpot, C. 1967. The differentiation of *Trichophyton mentagrophytes* from *T. rubrum* by a simple urease test. *Sabouraudia* 5:189-93.

Roberts, G. D., C. D. Horstmeier, G. A. Land, and J. H. Foxworth. 1978. Rapid urea broth test for yeasts. *J. Clin. Microbiol.* 7:584-88.

Zimmer, B. L. and G. D. Roberts. 1978. Rapid selective urease test for presumptive identification of *Cryptococcus neoformans. J. Clin. Microbiol.* 10:380-81.

Media and Reagent Preparation

APPENDIX B

1. Casein Agar

    A. Skim milk .... 10.0 gm
       Distilled water .... 90.0 ml
    B. Agar .... 3.0 gm
       Distilled water .... 97.0 ml

    Autoclave A and B separately at 121°C for 10 minutes. Cool to 50°C and combine A and B. Dispense 25 ml aliquots into sterile 15 × 100-mm Petri plates. Store at 4–8°C.

2. Cellobiose Medium

    $(NH_4)_2HPO_4$ .... 1.0 gm
    KCl .... 0.02 gm
    $MgSO_4 \cdot 7H_2O$ .... 0.2 gm
    Agar .... 15.0 gm
    Distilled water .... 1.0 l

    Place the above ingredients in a flask, heat to boiling with constant mixing, and then adjust pH to 7. Add 15 ml of a 0.04% solution of bromcresol purple. Dispense in 10-ml aliquots into screw capped tubes. Autoclave at 121°C for 15 minutes. Add 0.5 ml of a 10% aqueous filter-sterilized cellobiose solution to each tube. Allow the medium to solidify as slants, and store at 4–8°C.

3. Hypoxanthine Agar

    Beef extract .... 3.0 gm
    Peptone .... 5.0 gm
    Agar .... 15.0 gm
    Distilled water .... 1.0 l

    Place the above ingredients in a flask, heat to boiling with constant mixing, adjust pH to 7, and then dispense 100-ml aliquots into 250-ml flasks. Autoclave at 121°C for 10 minutes. Allow the medium to cool almost to solidification, and then add 0.5 gm of hypoxanthine. Mix well and dispense 25 ml aliquots into 15 × 100-mm Petri plates. Store at 4–8°C.

4. Lactophenol

    Phenol (concentrated) .... 20.0 ml
    Lactic acid .... 20.0 ml
    Glycerol .... 40.0 ml
    Distilled water .... 20.0 ml

    Place the above ingredients in a 250-ml flask and mix well. Store at room temperature.

5. Lactophenol Cotton Blue

    Dissolve 0.05 gm of cotton blue in the 20 ml of distilled water, then proceed as in the preparation of lactophenol.

6. Lysozyme Broth

    A. Basal glycerol broth
       Peptone .... 5.0 gm
       Beef extract .... 3.0 gm
       Glycerol .... 70.0 ml
       Distilled water .... 1.0 l

    Divide the solution into 2 equal portions; dispense one of the two portions into 5-ml quantities to be used as controls. Autoclave at 121°C for 15 minutes. Store at 4–8°C.

    B. Lysozyme broth
       1. Preparation of lysozyme solution
          Lysozyme .... 100.0 mg
          N/100 HCl .... 100.0 ml

    Sterilize by filtration

       2. Preparation of lysozyme broth
          Basal glycerol broth .... 95.0 ml
          Lysozyme solution .... 5.0 ml

    Mix and aseptically dispense in 5-ml aliquots. Store at 4–8°C.

7. Peptone Discs

    Peptone .... 3.0 gm
    Distilled water .... 10.0 ml

Dissolve the peptone in distilled water, and autoclave at 121°C for 15 minutes. Aseptically place 6-mm sterile paper discs in the solution and allow them to become saturated. Remove the discs and air dry in sterile Petri plates. Store at 4–8°C

8. Potassium Nitrate Discs

KNO_3	3.0 gm
Distilled water	10.0 ml

Proceed as for peptone discs (as in number 7 above).

9. Rice Grains

Unfortified white rice grains	8.0 gm
Distilled water	25.0 ml

Place the above ingredients in a 125-ml flask and autoclave at 121°C for 15 minutes. Store at 4–8°C.

10. Swab Nitrate Test Reagent

KNO_3	2.0 gm
$NaH_2PO_4 \cdot H_2O$	11.7 gm
Na_2HPO_4	1.14 gm
17% Zephiran Chloride solution	1.2 ml
Water	200 ml

Place the above ingredients in a flask and mix well. Adjust the pH to 5.8. Store at 5–8°C in a brown bottle until needed.

11. Tyrosine Agar

Proceed as in the preparation of hypoxanthine agar (as in number 3 above), except add 0.5 gm of tyrosine instead of hypoxanthine.

12. V-8 Juice Agar

V-8 juice	500.0 ml
Dry yeast	10.0 gm
Agar	20.0 gm
Distilled water	500.0 ml

Mix the V-8 juice and dry yeast together and adjust the pH to 6.8 with 20% potassium hydroxide. Dissolve the agar in the distilled water by heating. Mix the 2 solutions and dispense 10-ml aliquots into screw cap tubes. Autoclave at 121°C for 15 minutes. Allow the medium to solidify as slants, and then store at 4–8°C.

13. Wickerham's Yeast-Malt Agar

Yeast extract	3.0 gm
Malt extract	3.0 gm
Peptone	5.0 gm
Glucose	10.0 gm
Agar	20.0 gm
Distilled water	1.0 l

Place the above ingredients in a flask, heat to boiling with constant mixing, and dispense 10-ml aliquots into screw cap tubes. Autoclave at 121°C for 15 minutes. Allow the medium to solidify as slants and store at 4–8°C.

14. Xanthine Agar

Proceed as in the preparation of hypoxanthine agar (as in number 3 above) except add 0.4 gm of xanthine instead of hypoxanthine.

Selected Synonyms

APPENDIX C

Synonym	Accepted Name
Acrotheca	*Ramularia*
Allescheria boydii	*Pseudallescheria boydii*
Aspergillus glaucus	*Aspergillus glaucus* group
Basidiobolus haptosporus	*Basidiobolus ranarum*
Basidiobolus meristosporus	*Basidiobolus ranarum*
Cephalosporium	*Acremonium*
Cladosporium trichoides	*Cladosporium bantianum*
Cladosporium werneckii	*Exophiala werneckii*
Emmonsia	*Chrysosporium*
Emmonsiella	*Ajellomyces*
Entomophthora coronata	*Conidiobolus coronatus*
Histoplasma duboisii	*Histoplasma capsulatum* var. *duboisii*
Hormodendrum	*Cladosporium*
Keratinomyces	*Trichophyton*
Monosporium apiospermum	*Scedosporium apiospermum*
Mucor pusillus	*Rhizomucor pusillus*
Nocardia dassonvillei	*Nocardiopsis dassonvillei*
Petriellidium boydii	*Pseudallescheria boydii*
Phaeococcus	*Phaeococcomyces*
Phialophora compacta	*Fonsecaea compacta*
Phialophora dermatitidis	*Wangiella dermatitidis*
Phialophora gougerotii	*Exophiala jeanselmei*
Phialophora jeanselmei	*Exophiala jeanselmei*
Phialophora pedrosoi	*Fonsecaea pedrosoi*
Phialophora spinifera	*Exophiala spinifera*
Pityrosporum	*Malassezia*
Pullularia	*Aureobasidium*
Sporotrichum gougerotii	*Nomen dubium*
Sporotrichum schenckii	*Sporothrix schenckii*
Trichophyton gallinae	*Microsporum gallinae*
Trichophyton persicolor	*Microsporum persicolor*
Trichosporon cutaneum	*Trichosporon beigelii*
Trichosporon penicillatum	*Geotrichum penicillatum*
Zopfia rosatii	*Neotestudinia rosatii*

Index

Notes

Notes

Notes

Notes